Parasitic Wasps on Butterfly Expedition

Foraging Strategies of Egg and Larval Parasitoids Exploiting Infochemicals of Brussels Sprouts and Their *Pieris* Hosts

Dissertation
zur Erlangung des akademischen Grades des Doktors
der Naturwissenschaften (Dr. rer. nat.)

eingereicht im Fachbereich Biologie, Chemie, Pharmazie der
Freien Universität Berlin

angefertigt am Institut für Biologie
Angewandte Zoologie / Ökologie der Tiere

vorgelegt von

Nina Fatouros

aus Berlin

Berlin, im Juni 2006

Cover design by: Nina E. Fatouros

Photos by: Tibor Bukovinszky, Hans M. Smid, Nina E. Fatouros

Bibliografische Information Der Deutschen Bibliothek

Die Deutsche Bibliothek verzeichnet diese Publikation in der Deutschen
Nationalbibliografie; detaillierte bibliografische Daten sind im Internet über
http://dnb.ddb.de abrufbar.

ISBN 3-8325-1313-2

Logos Verlag Berlin

Comeniushof, Gubener Str. 47,

10243 Berlin

Tel.: +49 030 42 85 10 90

Fax: +49 030 42 85 10 92

INTERNET: http://www.logos-verlag.de

1. Gutachterin: Prof. Dr. Monika Hilker, Freie Universität Berlin

2. Gutachter: Prof. Dr. Marcel Dicke, Wageningen Universiteit, Niederlande

Disputation am 23. Juni 2006

To my German, Greek and Dutch family (and expected hybrids)...

This thesis is based on the following manuscripts:

1) **Fatouros, N.E.,** Dicke, M., Mumm, R., Meiners, T. & Hilker, M. (manuscript). Infochemical-exploiting strategies used by foraging egg parasitoids.

2) **Fatouros, N.E.,** Huigens, M.E., van Loon, J.J.A., Dicke, M., & Hilker, M. (2005a). Chemical communication - Butterfly anti-aphrodisiac lures parasitic wasps. *Nature*, **433**, 704.

3) **Fatouros, N.E.,** Bukovinszkine'Kiss, G., Kalkers, L.A., Soler Gamborena, R., Dicke, M., & Hilker, M. (2005b). Oviposition-induced plant cues: do they arrest *Trichogramma* wasps during host location? *Entomologia Experimentalis et Applicata*, **115**, 207-215.

4) **Fatouros, N.E.,** Zheng, S.-J., Müller, F., Burow, M., Dicke, M., & Hilker, M. (manuscript). Induced defense mechanisms in Brussels sprouts plants in response to *Pieris* egg depositions: chemical and gene expression analysis.

5) **Fatouros, N.E.,** Bukovinszkine'Kiss, G., van Loon, J.J.A., Huigens, M.E., Dicke, M., & Hilker, M. (manuscript). Butterfly anti-aphrodisiac involved in eliciting oviposition-induced plant responses.

6) **Fatouros, N.E.,** Bukovinszkine'Kiss, G., Dicke, M., & Hilker, H. The response specificity of *Trichogramma* egg parasitoids towards infochemicals during host location. Submitted to *Journal of Insect Behavior*.

7) **Fatouros, N.E.,** Van Loon, J.J.A., Hordijk, K.A., Smid, H.M., & Dicke, M. (2005c). Herbivore-induced plant volatiles mediate in-flight host discriminationby parasitoids. *Journal of Chemical Ecology*, **31**, 2033-2047.

Contents

1 | General Introduction and Thesis Outline

N.E.Fatouros

Chapter 1

General Introduction and Thesis Outline

Information conveyance through chemical signals is known to be the oldest mode of interaction between organisms (Bradburry and Vehrencamp 1998). For relatively small animals such as insects chemicals offer many advantages. Unlike other signals, chemical cues can work in the dark, can be perceived upon contact or at large distances, and additionally can persist in time. Therefore, insects use such so-called **infochemicals** (see Dicke and Sabelis 1988 for terminology) in almost every aspect of their life, like food finding, defense, or mate location (Berenbaum 1995).

A large class of infochemicals involved in chemical communication between animals are **pheromones** (from the Greek *pherein*, to carry or transfer, and *hormone*, to excite or stimulate). They are used between members of the same species, such as sex pheromones, which operate in mate finding and species recognition. Any pheromone or chemical produced by an organism can be exploited by other organisms, whether they are conspecifics, mutualists, or natural enemies (Wyatt 2003). Infochemicals can be classified into allomones, kairomones, or synomones depending on the cost and benefit to emitter and receiver (See table 1) (Nordlund 1981). Insect pheromones eavesdropped by natural enemies (then termed **kairomones**) and **synomones**, produced by plants and exploited by parasitoids in a mutualistic relationship comprise the largest part of chemical cues dealt with in this thesis.

The majority of **parasitoids** are parasitic wasps or flies of the order Hymenoptera or Diptera. There are about 70.000 described parasitoid species (of the total ca. 900.000 described insect species), but because of their inconspicuous life style many more parasitoid species are expected (Godfray 1994). Parasitoids are insects whose larvae live parasitic by developing on or in the bodies of other insects. However, parasitoids always kill their hosts and thus should be in this respect more considered as predators. After the parasitoid larvae have developed, the new generation of adult female parasitoids emerging has to locate new hosts for their offspring, whereas they themselves are mainly nectar feeders (Godfray 1994). This link between successful host location and offspring assurance puts the female

Table 1: Infochemicals classified by cost and benefit to emitter and receiver

Infochemical	Effect on emitter	Effect on receiver	Type communication
Allomone	+	-	Deceit, propaganda, defense
Kairomone*	-	+	Eavesdropping
Synomone	+	+	Mutualism

*Pheromones (infochemicals in intraspecific interactions) are termed as kairomones when they are exploited by natural enemies.
After Wyatt (2003), terminology after Dicke & Sabelis (1988).

parasitoid's searching behavior under strong selection pressure shown by the variety of cues used during host foraging (van Alphen and Vet 1986). The female often has to search for widely dispersed hosts, in or on which she lays her eggs (**endo- or ectoparasitoids** respectively). Most parasitoids are thus well-equipped to travel over larger distances and to locate their hosts by acoustic, visual, or chemical cues (Berenbaum 1995). Like parasites, parasitoids can be highly host-specific, mainly because those living as endoparasitoids must overcome the host's immune defense and additionally need to synchronize their development with that of their host.

Parasitoids can be classified by the host stage they attack into egg-, larval, pupal, or adult parasitoids. Furthermore, they can be divided into two groups on the ecological attributes of their host: larval parasitoids that allow hosts to continue to grow and metamorphose, and to be mobile after parasitism, are called **koinobionts**. The other group **idiobionts** are usually some larval parasitoids that kill/paralyze the host during oviposition or egg or pupal parasitoids whose hatching larvae start to consume the host subsequently (Askew and Shaw 1986).

A main incentive for studying the host searching behavior of parasitoids is to effectively utilize them as **biological control** agents (Lewis and Nordlund 1985). Since decades, natural enemies of economic pest species have been successfully used to reduce the population size of pests and to increase the yield with minimal application of chemical pesticides. *Trichogramma* parasitoids were used more than other natural enemies for biological control of Lepidopteran pests in various crops (van Lenteren 2000). Still, large gaps in our knowledge on parasitoids searching behavior remain after so many years of research. A key for the application of parasitoids in the field or in green houses and successful parasitization of the hosts is an intensive study of the parasitoids' behavior.

This thesis deals with the relationship of parasitic wasps, their herbivorous hosts, and their host plants, i.e. **tritrophic interactions**, and the chemicals responsible for information transfer between the different trophic levels. This concerns infochemical-exploiting strategies of parasitoids attacking host eggs and those searching for larval hosts. Infochemicals are known to be the major group of stimuli used in parasitoid-host interactions (Vinson 1976; Weseloh 1981). **Larval parasitoids** face the problem of mobile and often aggressive and defensive host larvae that are difficult to examine thoroughly for suitability. **Egg parasitoids** encounter different challenges like sessile and inconspicuous host eggs, which release much fewer cues than larval hosts do. Chapter 2 gives a comprehensive overview of different strategies used by egg parasitoids. The main difference with foraging strategies of larval parasitoids is their reliance on the adult host stage by spying on the host's intraspecific communication system sometimes in combination with phoresy (hitch-hiking on the transporting host) (See also chapter 3). Such an **infochemical detour** (Vet and Dicke 1992) via adult host pheromones is less common in larval parasitoids. *Leptopilina heterotoma*, a parasitoid that attacks larvae of *Drosophila* spp. is attracted to the aggregation pheromone of the adult *Drosophila* fly. Oviposition and larval hatches occur soon after the pheromone is placed on the substrate, thus this detour on an infochemical of the adult hosts seems a reliable solution for these larval parasitoids (Wiskerke et al. 1993). Yet, many larval parasitoids are known to use plant volatiles induced by larval feeding as long-range cues and host-derived cues in the closer vicinity (Geervliet et al. 1994; Steinberg et al. 1993; Tumlinson et al. 1993; Vet and Dicke 1992). Recently, egg parasitoids were shown to use plant volatiles induced by egg deposition of herbivorous insects: specialized egg parasitoids responded to volatiles of the pine tree, elm tree or a bean plant induced by eggs and wounding or feeding damage in the same moment (Colazza et al. 2004; Hilker et al. 2002; Meiners and Hilker 2000).

The **tritrophic system** investigated here consists of *Brassica oleracea* L. var. *gemmifera* (Brussels sprouts), the gregarious large cabbage white *Pieris brassicae* L. and the solitary small cabbage white *P. rapae* L. (Lepidoptera: Pieridae), and the parasitoids *Trichogramma brassicae* Bezdenko and *T. evanescens* Westwood (Hymenoptera: Trichogrammatidae) parasitizing *Pieris* eggs and *Cotesia glomerata* L. and *C. rubecula* Marshall (Hymenoptera: Braconidae) parasitizing *Pieris* caterpillars (Figure 1). Several aspects on parasitoid –host - host plant interactions in this system have been already studied. Brussels sprouts plants were shown to emit a blend of volatiles after being infested by *Pieris* caterpillars that attract *C. rubecula*, a solitary endoparasitoid of *P. rapae* and *C. glomerata*, a gregarious parasitoid of both *Pieris* species (Agelopoulos and Keller 1994; Geervliet et al. 1994; 1998; Mattiacci et al. 1995; 2001).

The elicitor responsible for the induction of plant volatiles, a β-glucosidase, was

Figure 1: The Brussels sprouts - *Pieris* system. (A) Brussels sprouts plant (Photo: NE Fatouros), (B) Ovipositing *Pieris brassicae* female (Photo: NE Fatouros), (C) *Trichogramma brassicae* wasp parasitizing *P. brassicae* eggs (Photo: NE Fatouros), (D) *Cotesia glomerata* wasp parasitizing *P. brassicae* caterpillars (Photo: HM Smid).

found in the regurgitant of *P. brassicae* (Mattiacci et al. 1995). Headspace volatiles from undamaged and *Pieris*-infested Brussels sprouts plants revealed differences in at least 20 compounds, which also elicited a response in the antennae of the *Cotesia* parasitoids in coupled gas chromatography-electroantennography (GC-EAG) analysis (Blaakmeer et al. 1994; Mattiacci et al. 1994; 1995; 2001; Smid et al. 2002). *T. evanescens* (later revised to *T. maidis* = *T. brassicae*) was shown to respond to volatile cues from virgin *P. brassicae* females as well as to contact cues of *Pieris* wing scales (Noldus and van Lenteren 1985a; Noldus and van Lenteren 1985b).

The **main goal** of this thesis is to discover new host foraging strategies of egg and larval parasitoids exploiting infochemicals of the first and second trophic level in the *Brassica-Pieris* system. More specifically, I want to address whether:

(1) Trichogramma egg parasitoids exploit the communication system of *P. brassicae* butterflies

(2) Trichogramma egg parasitoids make use of synomones of Brussels sprouts plants induced by *P. brassicae* eggs. These new aspects allow comparisons to be made to the three other systems, where indirect plant defense by herbivorous eggs was demonstrated and to the already shown induction induced by larval feeding (See table 2).

(3) Cotesia larval parasitoids make use of volatiles induced in Brussels sprouts plants by host feeding to discriminate between suitable and unsuitable, already parasitised hosts in-flight.

Chapter 2 provides a comprehensive overview of the different host location strategies of egg parasitoids with a focus on their infochemical use. Additionally, the infochemical-exploiting strategies of egg parasitoids are discussed in a tritrophic context also with respect to their learning abilities and their dietary breadth.

Chapter 3 provides an example of an infochemical-exploiting strategy of the host communication system by parasitoids: it was investigated whether *T. brassicae*

Table 2: Aspects planned to be elucidated in this research on the induction of indirect defense in Brussels sprouts by *Pieris brassicae* eggs: comparison between other investigated tritrophic systems concerning induction of indirect plant defense by insect egg deposition and investigations in the Brussels sprouts system on feeding-induced plant volatiles

Mechanism/ plant	Induction by oviposition		Induction by feeding
	trees/bean	cabbage	cabbage
	✔		✔
Volatile blend	By egg deposition + wounding/feeding	?	By larval feeding or regurgitant
Local	✔	?	✔
Systemic	✔	?	✔
Induction time	3 h – 5 d / 3 – 4 d	?	ca. 6 h
Elicitor	In oviduct secretion	?	In regurgitant

responds to specific volatiles of mated *P. brassicae* females and whether the egg parasitoid uses a so-called anti-sex-pheromone or anti-aphrodisiac to detect the female butterfly, and ride on them to the host plant to parasitize future laid egg clutches.

In **chapter 4**, olfactory and contact bioassays were conducted, to test whether egg deposition by *Pieris brassicae* induces Brussels sprouts plants to produce cues that attract or arrest *Trichogramma brassicae*. Additionally, different *P. brassicae* egg ages were tested for suitability.

Based on the results of the behavioral bioassays in chapter 3, a chemical analysis of leaf surfaces from egg-laden Brussels sprouts plants against uninfested plants was conducted in **chapter 5**, to elucidate the responsible chemical compounds of the egg parasitoids' arrestment. In addition, a total leaf analysis of glucosinolates, a secondary compound class characteristic for the Brassicaceae family, was conducted to see whether egg deposition induces a quantitative change in these compounds. Besides a chemical approach, I used a molecular one, by analyzing the expression of two genes encoding the enzymes myrosinase (MYR) and lipoxygenase (LOX), which are involved in the biosynthetic pathways of secondary defense plant compounds, to evaluate whether their transcription level in the plants changed after induction by *Pieris* eggs.

The next step, presented in **chapter 6**, was to elucidate whether the elicitor of the oviposition-induced plant synomone is present in the contents of the accessory gland of *P. brassicae* females, which is used to coat the eggs. The response of *T. brassicae* wasps towards plants treated with accessory glands against untreated plants was tested. Hereby, the possible role in this induction by egg deposition of male products transferred to the females during mating was of special interest. Furthermore, chemical analysis was conducted to examine possible differences in the composition between glands of virgin and mated *P. brassicae* females. In addition, female bursa copulatrix and male reproductive tissues were chemically analysed and compared between virgin and mated ones.

In **chapter 7**, the response specificity of the conspecific egg parasitoid *T. evanescens* was investigated towards a) plant cues induced by *P. brassicae* eggs and b) volatiles of the adult butterflies. Furthermore, their ability to mount the host and their preference for mated female hosts was elucidated. Finally, different host egg ages were tested for suitability.

Chapter 8 addresses the questions whether plant volatiles play a role in avoiding parasitoid competition by discriminating parasitized from unparasitized hosts in flight. Behavioral bioassays were conducted with two conspecific *Cotesia* species, to see whether they can discriminate between plants treated with unparasitized *Pieris* caterpillars or their regurgitant and plants treated with parasitized caterpillars or their regurgitant. Additionally chemical analysis of volatile blends of Brussels sprouts plants either treated with regurgitant of unparasitized or regurgitant of parasitized *P. brassicae* caterpillars were carried out.

Chapter 9 synthesizes the main results and conclusions of the first part of the thesis, the work on *Trichogramma* wasps with the second part on *Cotesia* host discrimination and compares them with general patterns of different strategies of larval and egg parasitoids. Furthermore, mechanisms and functions of plant synomones either induced by feeding or by oviposition are compared and discussed. Finally, directions for future research are suggested.

References

Agelopoulos NG, Keller MA (1994) Plant-natural enemy association in the tritrophic system, *Cotesia rubecula-Pieris rapae*-Brassicaceae (Cruciferae): II. Preference of *C. rubecula* for landing and searching. Journal of Chemical Ecology 20:1735-1748

Askew RR, Shaw MR (1986) Parasitoid communities: their size, structure and development. In: Waage J, Greathead D (eds) Insect Parasitoids. Academic Press, New York, pp 225-264

Berenbaum M (1995) Bugs in the System: Insects and Their Impact on Human Affairs. Addison-Wesley Publishing Company, Inc., Reading, Massachusetts

Blaakmeer A, Geervliet JBF, van Loon JJA, Posthumus MA, van Beek TA, de Groot Æ (1994) Comparative headspace analysis of cabbage plants damaged by two species of *Pieris* caterpillars: consequences for in-flight host location by *Cotesia* parasitoids. Entomologia Experimentalis et Applicata 73:175-182

Bradburry JW, Vehrencamp SL (1998) Chemical signals. In: Principles of Animal Communication. Sinaner Associates Inc. Publishers Sunderland, Massachusetts, pp 279-318

Colazza S, Fucarino A, Peri E, Salerno G, Conti E, Bin F (2004) Insect oviposition induces volatile emission in herbaceous plants that attracts egg parasitoids. Journal of Experimental Biology 207:47-53

Dicke M, Sabelis MW (1988) Infochemical terminology: based on cost-benefit analysis rather than origin of compounds? Functional Ecology 2:131-139

Geervliet JBF, Vet LEM, Dicke M (1994) Volatiles from damaged plants as major cues in long-rage host-searching by the specialist parasitoid *Cotesia rubecula*. Entomologia Experimentalis et Applicata 73:289-297

Geervliet JBF, Vreugdenhil AI, Dicke M, Vet LEM (1998) Learning to discriminate between infochemicals from different plant-host complexes by the parasitoids *Cotesia glomerata* and *C. rubecula*. Entomologia Experimentalis et Applicata 86:241-252

Godfray HCJ (1994) Parasitoids - Behavioral and Evolutionary Ecology. Princeton University Press, Princeton, New Jersey

Hilker M, Kobs C, Varama M, Schrank K (2002) Insect egg deposition induces *Pinus sylvestris* to attract egg parasitoids. Journal of Experimental Biology 205:455-461

Lewis WJ, Nordlund DA (1985) Behavior-modifying chemicals to enhance natural enemy effectiveness. In: Hoy MA, Herzog DC (eds) Biological Control in Agricultural IPM Systems. Academic Press, Orlando, pp 80-101

Mattiacci L, Dicke M, Posthumus MA (1994) Induction of parasitoid attracting synomone in Brussels sprouts plants by feeding of *Pieris brassicae* larvae: role of mechanical damage and herbivore elicitor. Journal of Chemical Ecology 20:2229-2247

Mattiacci L, Dicke M, Posthumus MA (1995) β-Glucosidase: An elicitor of the herbivore-induced plant odor that attracts host-searching parasitic wasps. Proceedings of the National Academy of Sciences of the United States of America 92:2036-2040

Mattiacci L, Rudelli S, Ambühl Rocca B, Genini S, Dorn S (2001) Systemically-induced response of cabbage plants against a specialist herbivore, *Pieris brassicae*. Chemoecology 11:167-173

Meiners T, Hilker M (2000) Induction of plant synomones by oviposition of a phytophagous insect. Journal of Chemical Ecology 26:221-232

Noldus LPJJ, van Lenteren JC (1985a) Kairomones for the egg parasite *Trichogramma evanescens* Westwood I. Effect of volatile substances released by two of its hosts, *Pieris brassicae* L. and *Mamestra brassicae*, L. Journal of Chemical Ecology 11:781-791

Noldus LPJJ, van Lenteren JC (1985b) Kairomones for the egg parasite *Trichogramma evanescens* Westwood II. Effect of contact chemicals produced by two of its hosts, *Pieris brassicae* L. and

Pieris rapae L. Journal of Chemical Ecology 11:793-800

Nordlund DA (1981) Semiochemicals: a review of the terminology. In: Nordlund DA, Jones RL, Lewis WJ (eds) Semiochemicals. Their Role in Pest Control. Wiley & Sons, New York, pp 13-28

Smid HM, Van Loon JJA, Posthumus M, A., Vet LEM (2002) GC-EAG-analysis of volatiles from Brussels sprouts plants damaged by two species of *Pieris* caterpillars: olfactory respective range of a specialist and generalist parasitoid wasp species. Chemoecology 12:169-176

Steinberg S, Dicke M, Vet LEM (1993) Relative importance of infochemicals from first and second trophic level in long-range host location by the larval parasitoid *Cotesia glomerata*. Journal of Chemical Ecology 19:47-59

Tumlinson JH, Lewis WJ, Vet LEM (1993) How parasitic wasps find their hosts. Scientific American 268:100-106

van Alphen JJM, Vet LEM (1986) An evolutionary approach to host finding and selection. In: Waage J, Greathead D (eds) Insect Parasitoids. Academic Press, London, pp 23-61

van Lenteren JC (2000) Success in biological control of arthropods by augmentation of natural enemies. In: Gurr GM, Wratten SD (eds) Biological Control: Measures of Success. Kluwer Academic Publishers, Hingham, pp 77-103

Vet LEM, Dicke M (1992) Ecology of infochemical use by natural enemies in a tritrophic context. Annual Review of Entomology 37:141-172

Vinson SB (1976) Host selection by insect parasitoids. Annual Review of Entomology 21:109-133

Weseloh RM (1981) Host location in parasitoids. In: Nordlund DA, Jones RL, Lewis WJ (eds) Semiochemicals: Their Role in Pest Control. John Wiley & sons, New York, pp 79-95

Wiskerke JSC, Dicke M, Vet LEM (1993) Larval parasitoid uses aggregation pheromone of adult hosts in foraging behaviour: a solution to the reliability-detectability problem. Oecologia 93:145-148

Wyatt TD (2003) Pheromones and Animal Behaviour: Communication by Taste and Smell. Cambridge University Press, Cambridge

2 | Infochemical-Exploiting Strategies Used by Foraging Egg Parasitoids

N.E. Fatouros, M. Dicke, R. Mumm, T. Meiners & M. Hilker

Chapter 2

Infochemical-Exploiting Strategies Used by Foraging Egg parasitoids

Abstract

During host location egg parasitoids rely on a variety of chemical cues originating from the adult host, host products, or the host plant rather than from the attacked host stage – the insect egg itself. Besides pupae, insect eggs are the most inconspicuous host stage attacked by parasitic wasps. To overcome the problem of low detectability of host eggs, egg parasitoids have developed several strategies such as exploiting long-range kairomones of the adult hosts, e.g. sex pheromones, plant synomones induced by egg deposition or host feeding or short-range contact cues derived from the adult host or the host plant. Moreover, egg parasitoids have developed the ability to use infochemical eavesdropping in combination with phoresy on the adult host to compensate their limited flight capability and to get access to freshly laid host eggs, a strategy that might be widespread among egg parasitoids.

Here, we provide a comprehensive overview of the different host location strategies of egg parasitoids with a focus on their infochemical use. Furthermore, the use of infochemicals by egg parasitoids is discussed with respect to their dietary breadth and their ability to learn.

Keywords: pheromones, chemical espionage, herbivore-induced plant volatiles, phoresy, dietary specialization

Introduction

Egg parasitoids, and in particular *Trichogramma* spp., are important insects used for biological control of insect pests worldwide (Li 1994; 1989; Romeis et al. 2005; van Lenteren 2000). By killing their host before it starts to damage the plant in the larval stage, egg parasitoids are likely to be more beneficial for the plant than larval parasitoids, where the parasitized host often continues to feed. Therefore, many efforts have been made in studying host selection behavior of egg parasitoids in order to optimize their application in biological control.

Generally, the process of host selection in parasitoids is divided into several behavioral phases: host habitat location, host location, and host acceptance. Host selection is mediated by numerous stimuli, of which chemicals are known to play the major role (Vinson 1976). Our review focuses on chemical cues used by egg parasitoids in the initial two stages of their host selection behavior before host contact takes place, i.e. long-range and short-range host location. Long-range cues used in habitat location are generally plant volatiles or herbivore pheromones, whereas cues used in the closer vicinity are mostly short-range cues of herbivore products or of the plant surface.

Vet and Dicke (1992) stated that herbivore-foraging carnivores face a reliability-detectability problem: infochemicals provided by the herbivore itself are reliable indicators of the herbivore's presence, but their detectability is low due to the small biomass of the herbivore relative to that of their host plant and due to selection on the host to minimize the emission. On the other hand, plant-derived stimuli are supposed to be more detectable because they are available in larger quantities and emitted over larger distances although they are not necessarily reliable indicators of the presence of the host. It was assumed that the evolution of host location is restricted by this reliability-detectability problem of prey/host-derived infochemicals. The authors proposed three solutions to this problem, namely a) using infochemicals produced by the adult host stage, such as pheromones to locate eggs or larvae of the herbivore (termed infochemical detour), b) using plant volatiles induced by host feeding (termed: herbivore-induced synomones, HIS, or herbivore-induced plant volatiles HIPV), or c) linking detectable cues with reliable, host-derived cues through associative learning.

From the egg parasitoids' point of view, plant volatiles induced by host feeding are indicating the presence of the host but not necessarily the presence of host eggs (infochemical detour). However, a new example of solution (b) is available for egg parasitoids: recently it was shown that egg parasitoids use plant volatiles induced by egg deposition (Colazza et al. 2004a; Hilker et al. 2002; for reviews see Hilker and Meiners 2002b; 2006; Meiners and Hilker 1997; 2000). These oviposition-induced plant cues were shown in laboratory bioassays to attract egg parasitoids, whereas feeding-induced plant volatiles in those species did not elicit any response. Thus, oviposition-induced cues are expected to be a more reliable solution for these egg parasitoids than plant cues induced by host feeding.

Egg parasitoids are specialized to develop in the eggs of other insects. Insect eggs are nutrient-rich enlarged cells that do not feed and produce feces and consequently lack intense long-range odors that can be exploited by their enemies. Additionally, insect eggs develop rapidly, except when they enter diapause, from a nutrition source to a complex embryo and might therefore be present only for 2 to 3 days (Hinton 1981). Many eggs are well protected either physically and/ or chemically (Hilker and Meiners 2002a) because of their immobility and their exposure to enemies and environmental influences. Some insects make their eggs less accessible by hiding them in plant tissue or by covering them with physical devices like hairs, feces, scales or secretion (Blum and Hilker 2002; Hilker 1994; Hinton 1981). Some chrysomelid beetles are known to use feces to protect their eggs (Hilker 1994). Yet, specialist parasitoids can use chemical cues from the feces to detect their host eggs (Hilker and Meiners 1999). Maternally incorporated or applied toxins can act as sufficient protection against their enemies (Bezzerides et al. 2004; Blum and Hilker 2002; Hilker and Meiners 1999). Consequently, egg parasitoids have to face these problems and encounter many challenges in finding their small and immobile hosts.

1 Long-range infochemicals

1.1 Plant cues

When an adult egg parasitoid emerges from its host egg, it may or may not find itself in the habitat which contains its preferred host. Consequently, egg parasitoids need to have an innate repertoire of responses to cues associated with the host habitat. Salt (1935) stated that most parasitoids are first attracted to a certain type of environment and then to a particular host. Vinson (1981) lists numerous factors involved in habitat location such as sound, visual cues, forms of electromagnetic radiation, and volatile chemicals. Of these factors chemical signals, mainly of plants, mostly provide cues to the host-habitat (Steidle and van Loon 2002; Vinson 1991; Weseloh 1981). These long-distance infochemicals usually arise from sources associated with or connected to the host stage attacked rather than from the host directly.

1.1.1   Volatiles from undamaged plants

A number of studies have shown *Trichogramma* spp. to respond to volatiles derived from undamaged plants or plant extracts (e.g. Altieri et al. 1982; Bar et al. 1979; Boo and Yang 1998; Kaiser et al. 1989; Nordlund et al. 1985; Reddy et al. 2002; Romeis et al. 2005). However, it seems that *Trichogramma* wasps and other egg parasitoids are generally arrested by plant volatiles after entering a habitat rather than attracted from a distance (Romeis et al. 1997). Due to their minute size egg parasitoids like *Trichogramma* spp. are known to disperse passively on wind drifts rather than to fly directly to an odor source (Noldus et al. 1991b). Wind speeds in the open field make directed upwind flight for *Trichogramma* impossible, which was also demonstrated in windtunnel experiments (Keller et al. 1985; Noldus et al. 1991b). However, within the vegetation either close to soil or within the canopy effects of wind might be different from those in open areas. It was shown that odor plumes disperse differently over open ground than in forests where turbulence and wind influences are less severe and odors spread over a smaller distance (reviewed by Murlis et al. 1992). Under low wind conditions *Trichogramma* spp. were shown to disperse in all directions almost uniformly (see Fournier and Boivin 2000 and references therein). The eulophid egg parasitoid *Oomyzus gallerucae* can be attracted by a synthetic plant compound to free-standing sticky traps (Meiners, pers. commun.), indicating the capacity to actively fly in the field.

Yet, the presence of nutrients (e.g. floral nectar or pollen) associated with the host plants or the host plant habitat may add to the attraction of the habitat (reviewed by Vinson 1991). Furthermore, these nutrients can increase the longevity and fecundity of the foraging female and may lead to increased parasitism (Romeis et al. 2005; Wäckers 2005; Winkler et al. 2005). *T. chilonis* females were arrested by volatiles from flowering sorghum plants (Romeis et al. 1997), whereas visual cues like white flowers were used by *T. carverae* during food location (Begum et al. 2004). In addition to indicating a food resource, flowers might also serve as a 'meeting point' where phoretic egg parasitoids climb on a nectar-feeding transporting host. This is known from phoretic mites and phoretic beetle larvae (Clausen 1976; Schwarz and Huck 1997). Phoresy is a means of dispersal successfully used by a number of scelionid and trichogrammid egg parasitoids and seen as an adaptation to their limited flight ability or to migrating hosts (see 1.2.3).

1.1.2   Herbivore-induced plant volatiles

Since nearly two decades it is known that plants can cooperate with the third trophic level - predators and parasitoids - to defend themselves indirectly against herbivore infestation. Various studies demonstrated that plants respond to feeding herbivores by induction of volatiles that attract the enemies of the herbivores (reviewed by Dicke et al. 1990; Dicke and van Loon 2000; Dicke and Vet 1999; Turlings et al. 2002). However, plants are also able to notice insect oviposition and to change their release of volatiles in response to egg deposition. Egg parasitoids have been shown to use such oviposition-induced plant synomones to locate their host eggs. Thus, insect egg deposition may induce "early alert" in a plant, prior to feeding damage by larvae (Hilker et al. 2002; Hilker and Meiners 2002b; 2006).

Meiners and Hilker (1997; 2000) were the first to demonstrate that plants are able to respond to egg deposition by a herbivorous insect in such a way that plant volatiles are released attracting egg parasitoids. They showed this in the field elm (*Ulmus minor*) that produces volatiles induced by oviposition of the elm leaf beetle *Xanthogaleruca luteola*. These oviposition-induced elm leaf volatiles attract the egg parasitoid *O. gallerucae*. The elm leaf beetle removes the epidermis of the leaf before egg deposition. However, the removal of plant epidermal tissue *per se* does not induce the volatiles attracting *O. gallerucae* nor do the eggs release attractants, nor does a combination of egg odor and plant volatiles attract the egg parasitoids. Indeed, the egg deposition induces the release of volatiles from elm leaves. Nevertheless, the removal of plant tissue is required for the volatile induction, as no attractive volatiles are emitted when eggs are transferred to intact leaves. The elicitor of the plant's response to egg deposition was found to be present in the oviduct secretion that envelopes the eggs and glues them to the leaves (Meiners and Hilker 2000).

Similar results were found in another tritrophic system where the eulophid wasp, *Chrysonotomyia ruforum*, is attracted by volatiles of the Scots pine (*Pinus sylvestris*) induced by eggs of the sawfly *Diprion pini* (Hilker et al. 2002; Mumm and Hilker 2005). The sawfly does not only remove the plant epidermis when ovipositing, but slits the pine needle longitudinally to insert eggs in a row. Again the elicitor was found to be present in the oviduct secretion coating the eggs. Here, strong indication is available that the elicitor is a small protein (Hilker et al. 2005).

In both the elm and pine system a specific response by specialized egg parasitoids toward egg-infested plants was demonstrated: neither the eggs alone nor artificially damaged plants mimicking the oviposition-related wounding were arresting the egg parasitoids. The plant responses were both local and systemic: egg-laden plant parts as well as uninfested plant parts near parts with eggs emitted attractive volatiles (Hilker et al. 2002; 2005; Meiners and Hilker 2000).

When analyzing the odor blends of the plants after egg deposition, changes of the quantity and quality of the volatile terpenoids have been detected for egg-laden elm leaves (Wegener and Schulz 2002; Wegener et al. 2001), whereas only quantitative changes of the terpenoid pattern have been found in pine needles. For example, pine twigs with eggs emitted the sesquiterpene (E)-β-farnesene in higher quantities than egg-free pine twigs (Mumm et al. 2003). The egg parasitoid *C. ruforum* uses the ratio of this single compound within a complex pine volatile background to detect its host eggs (Mumm and Hilker 2005).

In a system studied by Colazza et al. (2004a; b), not only egg deposition, but also feeding of the adults is necessary to induce volatiles attracting an egg parasitoid. It was shown that a generalist egg parasitoid, *Trissolcus basalis* responds to plant synomones emitted by the leguminous plant *Vicia faba* and induced by the pentatomid bug *Nezara viridula*. Attraction took place only when egg deposition occurred together with plant damage, in this case feeding activity of *N. viridula*. However, the eggs were laid on the undamaged plant part. Chemical analysis revealed that feeding together with oviposition of the pentatomid bugs induced the *V. faba* plants to produce blends of volatiles characterized especially by increased amounts of (E)-β-caryophyllene (Colazza et al. 2004b).

In all three mentioned systems, the egg-infested plant only emits the specific volatile blend when simultaneously damaged by either deposition of eggs (elm leaf beetle and pine sawfly) or feeding by adults (pentatomid bug). Additionally, mainly the terpenoid pattern in the odor blend seems to be modified after induction.

Other studies demonstrated that also feeding damage of a plant alone (without egg deposition) is able to induce volatiles attracting egg parasitoids. Moraes et al. (2005) have shown in a tritrophic system consisting of a leguminous plant – brown stink bug (*Euschistus heros*) –*Telenomus podisi* that plant volatiles induced by the feeding of adult *E. heros* alone were attractive to the egg parasitoid *Te. podisi*. However, egg-infested plants were not tested by these authors. Also *Anagrus nilaparvatae*, an egg parasitoid of the rice brown plant hopper (*Nilaparvata lugens*), is attracted to volatiles released from feeding-damaged host plants, in particular when they are infested by adult females or host nymphs (Lou et al. 2005). The wasps were not attracted to volatiles from female adults, eggs or other direct host cues. The authors suggested that the response to feeding-induced volatiles is adaptive to the egg parasitoids. All developmental stages can co-occur on the same plant, which makes an indirect association of host adult or nymph feeding with egg presence a reliable cue. Again, it was not tested whether egg deposition on the rice plant induces volatiles the attract egg parasitoids.

1.2 Host cues

Within a (micro-)habitat egg parasitoids are able to distinguish infested from uninfested areas. This phase of their foraging behavior can be mediated by cues originating from adult host insects. Spying on the host communication system seems a common strategy in predatory and parasitic arthropods (Stowe et al. 1995; Vinson 1984; Zuk and Kolluru 1998). Several signaling modalities, including acoustic, visual, and olfactory information channels are known to be exploited. Intraspecific cues are often host specific and thus reliable information for eavesdropping enemies and can additionally be detected over some distances. Intraspecific chemical cues used by parasitoids can either directly or indirectly reveal the presence of the preferred host stage (e.g. Godfray 1994; Hilker and Meiners 1999; Jones et al. 1973; Nordlund 1994; Steidle and van Loon 2002; Vinson 1976; 1985; 1991; Weseloh 1981). Egg parasitoids are known to use mainly sexual signals of the adults such as sex pheromones, sometimes in combination with phoresy.

1.2.1 Exploitation of sexual signals

Insects producing mating signals often face a trade-off between the attraction of the mate and eavesdropping by enemies. Long-range olfactory sex signals are usually emitted by females. Visual or acoustic signals are almost always produced by males as well as pheromones usually functioning at close-range as part of the courtship activity (Greenfield 1981; Thornhill 1979). More and more studies demonstrate chemical espionage by egg parasitoids of sexual host signals of moths (e.g. Arakaki et al. 1996; 1997; Boo and Yang 2000; Frenoy et al. 1992; Lewis et al. 1982; Noldus 1988; Noldus et al. 1990; 1991b; Nordlund et al. 1983; Reddy et al. 2002; Schöller and Prozell 2002), butterflies (Fatouros et al. 2005b; Noldus and van Lenteren 1985a) or sawflies (Hilker et al. 2000). Studies with *Trichogramma pretiosum* showed for the first time that an egg parasitoid uses volatile host pheromones for location of eggs. The presence of chemicals emanating from an excretion and volatiles of the abdominal tips of female *Heliothis zea* moths increased parasitization rates by *T. pretiosum*. Additionally, the presence of the synthetic chemicals, consisting of hexadecanal, (Z)-7-hexadecanal, (Z)-9-hexadecanal and (Z)-11-hexadecanal, identified as the sex pheromone of *Heliothis zea* females increased parasitization of *H. zea* eggs (Lewis et al. 1982). Later, laboratory studies confirmed the response of *T. pretiosum* to calling *H. zea* females (Noldus 1988).

Subsequently, this evidence received considerable attention and revealed a large impact of female sex pheromones on the host location behavior of egg parasitoids helping them to reach potential oviposition sites or to move within a plot. In about 75 % of the analyzed literature, the tested egg parasitoid species responded to odors of virgin female hosts or even the synthetic sex pheromone whereby in most cases a moth was tested as an odor source (Table 1). Powell (1999) gives a detailed overview on the responses of egg parasitoids to sex pheromones from lepidopteran hosts.

However, some authors suggest that host sex pheromones rather arrest egg parasitoids than attract them to a new habitat (Noldus et al. 1991b; Powell 1999). Many studies showed an intensive searching behavior of *Trichogramma* spp. in areas with these cues and an increase of parasitization (e.g. Boo and Yang 2000; Lewis et al. 1982; McGregor and Henderson 1998; Noldus 1988; Noldus and van Lenteren 1985a; Noldus et al. 1991b). Others showed that egg parasitoids are able to use host pheromones as an attractant and also forage directly for eggs in areas in which they perceive it (Aldrich et al. 1984; Bruni et al. 2000).

Besides *Trichogramma* species, other egg parasitoids like the scelionid *Telenomus* spp. are known to respond to moth sex pheromones. *Te. isis* was recently shown to be arrested by calling females of the African pink stem borer *Sesamia calamistis* (Fiaboe et al. 2003), and *Te. busseolae* responded to calling *S. calamistis* and the sex pheromone of *S. nonagrioides* (Colazza et al. 1997; Fiaboe et al. 2003). Non-calling moths did not arrest the scelionids. However, a response of egg parasitoids to calling virgin female moths does not necessarily mean a response to the host sex pheromone. In some cases, the sex pheromone blend or main compounds of the virgin moths elicited no response in the tested *Trichogramma* spp. (Frenoy et al. 1992; Noldus et al. 1990; Noldus and van Lenteren 1985a; Renou et al. 1992). Here, the kairomone seems to be a compound associated with the pheromone production rather than the pheromone itself.

Not only sex pheromones of moths, but also of sawflies can lure egg parasitoids. The two eulophid wasps *Chrysonotomyia ruforum* and *Dipriocampe diprioni* were arrested to the major sex pheromone compounds of their sawfly hosts *Diprion pini* [acetate and propionate of (2S,3R,7R)-3,7-dimethyl-2-tridecanol] and *Neodiprion sertifer* [(2S,3S,7S)-3,7-dimethyl-2-pentadecyl acetate]. Interestingly, *C. ruforum* responded only to the intraspecifically active stereoisomer of the pheromone of *N. sertifer*. The tested *C. ruforum* wasps originated from allopatric populations and different host species. The phoretic *Telenomus euproctidis* occurs on two allopatrically occurring tussock moths. Local wasp strains discriminated between the sex pheromone of their locally occurring moth, *Euproctis pseudoconspera* (10,14-dimethylpentadecyl isobutyrate) and the sex pheromone of *E. taiwana* [(Z)-16-methyl-9-heptadecenyl isobutyrate] (Arakaki et al. 1997). However, in the case of the sawfly parasitoid *C. ruforum* no population differences of the wasps' response towards the sex pheromones were found (Hilker et al. 2000).

Noldus et al. (1991a) showed that the sex pheromone of the female *Mamestra brassicae* adsorbs onto the leaf surface of cabbage plants to such an extent that foliage exposed to the pheromone attracts conspecific males but simultaneously arrested *Trichogramma evanescens* females. By this strategy the diurnally foraging egg parasitoids are able to use kairomones still detectable after some hours when being released by nocturnal moths. It was suggested by the authors that this synergism between host plant and host sex pheromone may act as a 'bridge in time' for the chemical spies. This could be a common phenomenon since many hosts where sex pheromones are exploited by an egg parasitoid are nocturnal moths and the pheromone compounds are all long-chain fatty acid derivatives with similar chemical properties.

Arakaki and Wakamura (2000) discovered a different 'bridge in time' strategy: *Telenomus euproctidis*, a phoretic egg parasitoid of the tussock moth *Euproctis taiwana* becomes active during the day, whereas its host is active during dawn. However, *Te. euproctidis* wasps were still able to detect trace amounts of the host sex pheromone even 48 hours after release, being retained on scale hairs at the anal tufts of female moths.

Few studies indicated an attraction of egg parasitoids to mated host females. Olfactometer studies showed that females of the egg parasitoid *Trissolcus basalis* responded to volatiles from virgin males of its host *Nezara viridula*, but also to mated females in a preovipositional state. Virgin females were not attractive to the parasitoids (Colazza et al. 1999). On the other hand, *Tr. brochymenae* wasps were attracted to volatile chemicals of virgin and mated females of its host *Murgantia histrionica*. However, their response to mated host females seemed to be particularly intense (Conti et al. 2003). In sexually reproducing species, mated females might provide more reliable cues for host searching in egg parasitoids than virgin ones since mated ones are closer to egg deposition. Indeed, a host anti-sex pheromone, an anti-aphrodisiac transferred by the male to the female during mating, has recently been shown to be exploited by *Trichogramma brassicae* females (Fatouros et al. 2005b) (Chapter 3) (see 1.2.3).

Bai et al. (2004) found an olfactory response of *Trichogramma ostriniae* towards mated females of the Asian corn borer (*Ostrinia furnacalis*) before their first oviposition and a response to their accessory glands. However, virgin females, mated females after their first oviposition, or their accessory glands did not elicit any response. Still, an airborne chemical from the egg masses, (*E*)-12-tetradecenyl acetate, which is the main component of the Asian corn borer sex pheromone, arrested the egg parasitoids as well. In *Pieris napi* egg deposition is followed by a reduction in the titre of an anti-aphrodisiac pheromone (Andersson et al. 2004). Such a post-mating female odor could also play a role in the Asian corn borer luring *T. ostriniae* to just-mated host females but not to females that had already oviposited. The latter might have a decrease in their anti-sex odor titre, which makes them less detectable to the parasitoids. Yet, it seems not clear whether the sex pheromone plays a significant role for *T. ostriniae* to locate the moths. Nevertheless, the exploitation of anti-aphrodisiacs or similar volatile pheromones of mated host females by egg parasitoids could be more widespread than so far assumed.

Eavesdropping on the host's sexual communication is supposed to be a stable strategy since the host is probably under strong selection pressure not to change the chemistry of its intraspecific signals. However, the detectability of sexual signals can be temporally restricted. Additionally they can be emitted away from oviposition sites, which limits the risk of being exploited, and this limits their reliability for egg parasitoids (Steidle and van Loon 2002). Unlike with anti-aphrodisiacs of mated host females, exploiting the sex pheromone of virgin females does not ensure the egg parasitoid any access to host eggs, because the host female still has to find a mate before it can oviposit.

Except for phoretic egg parasitoids many studies still show that the use of sexual signals can be beneficial for egg parasitoids: plants sprayed with synthetic pheromones arrest egg parasitoids, stimulate intensive searching behavior in these areas, and increase the probability to encounter host eggs resulting in high parasitism rates (Lewis et al. 1982; Nordlund et al. 1983).

1.2.2 Exploitation of hemipteran pheromones

In some Heteroptera, male adults emit a pheromone that attracts adults of both sexes (Aldrich 1995). Mating and oviposition often occur in the same area where these aggregation pheromones are emitted, which makes them a potential host location cue for egg parasitoids. *Gryon pennsylvanicum*, an egg parasitoid of the leaf-footed plant bug (*Leptoglossus australis*), was captured by traps baited with male bugs. Parasitization was increased when cages with male bugs were placed next to host eggs (Yasuda and Tsurumachi 1995).

Telenomus calvus is a specialized phoretic egg parasitoid of *Podisus maculiventris* and *P. neglectus*. It is known to utilize the aggregation pheromone of the male bugs, α-terpineol, to come in the host male's vicinity and to wait for females to arrive, whereupon they become phoretic after mating (Aldrich 1995; Aldrich et al. 1984; Buschman and Whitcomb 1980). Some females of the non-phoretic more generalistic *Te. podisi* were caught in male-pheromone baited traps, too (Aldrich 1985), but did not discriminate between host eggs placed in traps baited with the synthetic male pheromone of *P. maculiventris* or unbaited traps (Bruni et al. 2000). This implies that the synthetic pheromone does not affect the searching behavior of the non-phoretic *Te. podisi* females like it does in the phoretic *Te. calvus*. It was assumed that *Te. calvus* relies on other cues such as infochemicals of the eggs or from the host substrate.

1.2.3 Host location by phoresy in combination with chemical espionage

Phoresy is the transport of organisms on the bodies of others for purposes other than direct parasitism (Howard 1927). Phoresy in parasitoids is almost exclusively reported in egg parasitoids. Most phoretic egg parasitoids belong to the hymenopteran family Scelionidae and in some species of Trichogrammatidae, which are known to be phoretic on moths, grasshoppers, pentatomid bugs, dragonflies, and others. Clausen (1976) has given a detailed overview on this phenomenon. Phoretic egg parasitoids first have to find their transporting host, however, their host location cues are in most cases unknown (Table 2). A number of grasshoppers have been reported to be attacked by phoretic scelionids. By being phoretic it provides them access to large numbers of eggs oviposited in widely dispersed locations such as occurs with grasshoppers or to eggs sealed in containers such as occurs with mantids.

A fascinating, almost obligatory phoretic relationship is known of the scelionid wasp, *Mantibaria manticida*, that parasitizes the eggs of the European mantis (*Mantis religiosa*). The female wasp removes her wings after having mounted the host female, and hitches a ride with the mantis to the oviposition site. By being wingless the wasp is able to enter the frothy coating of the freshly laid mantis eggs. Afterwards the parasitoid often returns to the same female mantis to parasitize successive egg masses (Couturier 1941).

The exact stimulus that draws the female parasitoid to the host female is known only in a few cases. Arakaki et al. (1995; 1996) showed that the egg parasitoid *Telenomus euproctidis* is attracted by the sex pheromone [(Z)-16-methyl-9-heptadecenyl isobutyrate] of virgin tussock moths (*Euproctis pseudoconspersa*) before it mounts them, hiding in the anal tuft until the female moth starts laying her egg clutch. *Te. calvus* wasps utilize the aggregation pheromone of the male spined soldier bugs to become phoretic on the latter. After mating the wasp mounts the female, which is possibly recognized by female-specific volatiles from small glands underneath the wings (Aldrich 1995; Aldrich et al. 1984; Bruni et al. 2000; Buschman and Whitcomb 1980). Orr et al. (1986) reported a number of behavioral and biological adaptations of *Te. calvus* to its phoretic behavior. Because of the telescopic but rather weak ovipositor, known from scelionids (reviewed by Quicke 1997), *Te. calvus* is restricted to soft and young host egg masses less than 12 hours old. A phoretic habit ensures the availability of freshly laid eggs and could be one explanation why it has evolved quite frequently among the scelionids.

Recently, Fatouros et al. (2005b) (Chapter 3) reported about a chemical spy, *Trichogramma brassicae*, utilizing the anti-aphrodisiac benzyl cyanide, to become phoretic on its host, the large cabbage white (*Pieris brassicae*). The anti-aphrodisiac is transferred from males to females during mating to render the females less attractive to conspecific mates (Andersson et al. 2003). Mated females were most attractive to the wasps, unlike in most other cases of chemical espionage where virgin females lured egg parasitoids To examine the importance of the anti-aphrodisiac for phoresy in *T. brassicae*, the nitrile benzyl cyanide was applied onto virgin females. Indeed, the application of the anti-aphrodisiac rendered the virgin females attractive to the wasps and stimulated them to mount their hosts. Windtunnel experiments demonstrated that the wasps were able to hitch-hike on the butterflies until these landed on their host plant, then dismounted the host, and parasitized the freshly laid host eggs.

In general, we expect phoresy to be more adaptive to egg parasitoids with hosts laying eggs in clutches. Most phoretic egg parasitoids produce a small amount of eggs; in this way the female wasp can get rid of most eggs with the first encountered host egg clutch but also risks to loose her eggs in a less suitable host. Table 2 shows the reproductive biology of those hosts, known to be attacked by phoretic egg parasitoids: indeed, more than 90 % of the species are known to be gregarious hosts. Phoresy in combination with chemical espionage might be more widespread among egg parasitoids than assumed so far (Fatouros et al. 2005b) (Chapter 3).

2   Close range and contact infochemicals

2.1   Plant cues

After landing on the host plant, egg parasitoids mainly use contact or short-range volatiles that indicate the presence of the host or more specifically its eggs. Studies by Fatouros et al. (2005a) (Chapter 4) indicate that chemical changes in the leaf surface of Brussels sprout plants induced by eggs of *Pieris brassicae* arrest *Trichogramma brassicae*. However, no indication was found that egg deposition of *P. brassicae* induces volatiles that are used by the egg parasitoid for host location. Unlike in the other systems where egg deposition induces volatile plant cues (see 1.1), the *Pieris* butterfly is not severely damaging the leaf of a Brussels sprouts plant before or during the oviposition. The response of the plant seems restricted to leaf surface modifications. It is assumed that chemical changes of the pattern of compounds in the wax layer are induced that guide the egg parasitoids to suitable host eggs after landing on the host plant. The elicitor of this oviposition-induced synomone was shown to be located in the accessory gland secretion of mated but not of virgin *P. brassicae* females (Chapter 6).

The oviduct secretion or accessory gland secretion of herbivorous insects is used to envelope and/or glue the eggs onto the leaves (Gillott 2002; Hilker and Meiners 2006). Egg washes of *P. brassicae* were shown to deter the oviposition of conspecific butterflies (Klijnstra 1986). It was assumed that an oviposition-deterring pheromone, consisting of alkaloid miriamids was secreted onto their eggs (Blaakmeer et al. 1994b), and that this chemical marker arrested the egg parasitoid *T. evanescens* (Noldus and van Lenteren 1985b). However, later evidence accumulated that the cabbage leaves themselves produce oviposition deterrents in response to egg batches, because the pheromone was not detected in leaves from which eggs had been removed two days after deposition and that still showed deterrence (Blaakmeer et al. 1994a). This supports the hypothesis that the Brussels sprouts plants respond to egg deposition of *P. brassicae* by changing their chemical leaf surface profile, which deters the herbivores and arrests natural enemies at the same time.

2.2   Host cues

Laing (1937) was the first to report an arrestment behavior of *Trichogramma evanescens* wasps towards chemical traces left by the grain moth *Sitotroga cerealella*. Later, Lewis et al. (1971) found that cues left by ovipositing *Heliothis zea* or *Plodia interpunctella* moths increased parasitism by *T. evanescens*. Wing scales of the moths were the kairomonal source (Lewis et al. 1972). Several follow-up studies showed a response of *Trichogramma* wasps to host scales (e.g. Fatouros et al. 2005a; Noldus and van Lenteren 1985b; Shu and Jones 1989; Thomson and Stinner 1990; Zaborski et al. 1987). Jones et al. (1973) found hydrocarbons like tricosane in the scales of *Heliothis zea* as the biologically active cues, eliciting significant orientation and stimulation of parasitism by *T. evanescens*. Similar findings on other *Trichogramma* spp. using hydrocarbons were published by Lewis et al. (1975), Shu et al. (1990), Padmavathi and Paul (1998), and Paul et al. (2002).

Earlier investigations revealed that fresh accessory reproductive gland products have a kairomonal effect on egg parasitoids. Noldus and van Lenteren (1985) demonstrated an arresting effect of fresh extracts of *Pieris brassicae* egg washes to *Trichogramma evanescens*. Furthermore, it was shown for other egg parasitoids that kairomones from the host accessory gland or oviduct secretion played a role during host location or host recognition (Bai et al. 2004; Bin et al. 1993; Leonard et al. 1987; Nordlund et al. 1987; Strand and Vinson 1982; Strand and Vinson 1983). The fresh accessory glandular secretion of mated female *P. brassicae* arrested *T. brassicae* wasps, but also contains the elicitor for the induction of leaf surfaces changes. The induced leaf surface changes did not become effective one day after contact of the secretion with the leaf, but three days later *T. brassicae* was arrested on leaf area in vicinity to the site where secretion had been applied (Chapter 6).

Chemical traces of the European corn borer (*Ostrinia nubilalis*) applied onto the corn leaf surface in the vicinity of egg masses arrested *T. brassicae* wasps and increased parasitization (Garnier-Geoffroy et al. 1996). Such chemical trails might originate from the accessory reproductive gland as well or could be residues from a (host marking) pheromone applied on the leaf surface.

Fecal volatiles of both larvae and adults of the elm leaf beetle (*Xanthogaleruca luteola*) were attractive to its egg parasitoid *Oomyzus gallerucae* (Meiners and Hilker 1997). However, neither eggs, gravid females nor the larvae themselves were attractive. A substrate contaminated with feces of chrysomelids also arrested two eulophid egg parasitoid species after antennal contact took place (Meiners and Hilker 1997; Meiners et al. 1997).

In some cases, airborne chemicals of the host eggs themselves were shown to stimulate an intensive search behavior (Frenoy et al. 1992; Renou et al. 1992). Analysis of the eggs of *Ostrinia nubilalis* and *Mamestra brassicae* revealed the presence of various compounds on the egg surface such as hydrocarbons being responsible for the kairomonal activity on *Trichogramma brassicae* (Renou et al. 1989; 1992). Yet, other *Trichogramma* spp. did not react to host eggs from a short distance (Kaiser et al. 1989; Laing 1937). The eulophid egg parasitoid *Edovum puttleri* responded to egg washes of the Colorado potato beetle (*Leptinotarsa decemlineata*) with an increased searching time (Leonard et al. 1987). Kairomones of the egg surface often play a role in host recognition. In general, it is not expected that host eggs elicit long-range attraction by chemicals in egg parasitoids.

3 Infochemical use in a tritrophic context

As presented above, the foraging behavior of egg parasitoids is strongly affected by chemical cues from the first and the second trophic level. Vet and Dicke (1992) offered a conceptual framework by placing the ecology and evolution of infochemical use by herbivore-foraging carnivores in a tritrophic context. The central problem of foraging carnivores is the reliability-detectability problem of infochemicals. Vet and Dicke (1992) mentioned three solutions for this problem, which we discuss with respect to egg parasitoids below (See also introduction). Furthermore, they proposed some general rules that predict how parasitoids solve the reliability-detectability problem. Parasitoids specialized at the host and host plant level should use different foraging strategies and thus different infochemicals than those parasitizing numerous hosts, which are polyphagous herbivores themselves. Thus, the parasitoids behavioral response to a cue strongly depends on their degree of dietary specialization but also on the dietary specialization of their host species. The used infochemicals can be either specific for a certain host complex or generally present in various complexes and the reaction to the cues can be either innate or learned.

Table 1: Responses of egg parasitoids to volatile kairomones of adult hosts

Egg parasitoid	Host	Source	Response	Reference
Scelionidae				
Trissolcus brochymenae	*Murgantia histrionica*	Virgin & mated females Males	Attraction in Y-tube olfactometer	Conti et al. (2003)
Tr. basalis	*Nezara viridula*	Mated females Virgin males Defensive methathoracic gland scent	Attraction in (Y-tube) olfactometer	Colazza et al. (1999) Mattiacci et al. (1993)
Telenomus remus	*Spodoptera frugiperda*	Virgin female abdominal tips Synthetic sex pheromone	Attraction in Y-tube olfactometer Increased parasitism rates	Nordlund et al. (1983)
Te. euproctidis	*Euproctidis taiwana*	Virgin females	Attraction in the field Phoresy	Arakaki et al. (1996)
Te. calvus	*Podisius maculiventris*	Males Synthetic aggregation pheromone Mated females	Attraction in the field Phoresy	Bruni et al. (2000) Aldrich et al. (1984), Aldrich (1995)
Te. isis	*Sesamia calamistis*	Calling virgin females	Arrestment in 4-arm olfactometer	Fiaboe et al. (2003)
Te. busseolae	*Sesamia nonagrioides*	Calling virgin females	Attraction in Y-tube olfactometer	Colazza et al. (1997)
	Sesamia calamistis	Calling virgin females	Arrestment in 4-arm olfactometer	Fiaboe et al. (2003)
Ooencyrtus p.tyocampae	*Thaumetopoea pityocampa*	Virgin females	Attraction in the field	Battisti (1989)
Eulophidae				
Chrysonotomyia ruforum	*Diprion pini* *Neodiprion sertifer*	Synthetic sex pheromone	Arrestment in 4-arm olfactometer	Hilker et al. (2000)
Dipriocampe derrioni	*Diprion pini*			
Entodon leucogramma	*Scolytus multistriatus*	Synthetic aggregation pheromone	Attraction in the field	Kennedy (1984)

Gyron pennsylvanicum	*Leptoglossus australis*	Males	Attraction in the field Increased parasitism	Yasuda & Tsurumachi (1995)
Encyrtidae				
Oencyrtus nezarae	*Riptortus clavatus*	Synthetic aggregation pheromone	Attraction in the field	Leal et al. (1995)
Trichogrammatidae				
Lathromeris ovicida	*Sesamia calamistis*	Calling virgin females	Arrestment in 4-arm olfactometer	Fiaboe et al. (2003)
Uscana lariophaga	*Callosobruchus maculates*	Virgin females	Attraction	van Huis et al. (1994)
Trichogramma evanescens	*Ephestia spp.* *Plodia interpunctella*	Synthetic sex pheromone	Arrestment in 4-arm olfactometer	Schöller and Prozell (2002)
	Pectinophora gossypiella *Spodoptera littoralis* *Earias insulana*	Synthetic components of sex pheromone	Attraction in olfactometer	Zaki (1985)
T. brassicae (or T. maidis)	*Pieris brassicae*	Mated female butterflies Synthetic anti-sex pheromone	Arrestment in 2-chamber olfactometer Phoresy	Fatouros et al. (2005b)
	Ostrinia nubilalis	Calling virgin females No response to main sex pheromone component Eggs + pheromonal blend + maize extract	Attraction in linear olfactometer	Frenoy et al. (1992) Kaiser et al. (1989)
	Lobesia botrana	Calling virgin females	Slide increase in movement Increased parasitism	Garnier-Geoffroy et al. (1999)
		Calling virgin females	Arrestment in 4-arm-olfactometer	Noldus (1988)
	Plutella xylostella	Synthetic compound of sex pheromone	Increased parasitism	Klemm and Schmutterer (1993)

T. sibericum	*Rhodobota naevana*	Main component of sex pheromone	Increased searching time, arrestment	McGregor and Henderson (1998)
T. ostriniae	*Ostrinia furnacalis*	Mated female moths (before oviposition) ARGs of mated females	Attracted in olfactory bioassays	Bai et al. (2004)
T. oleae	*Palpita unionalis*	Synthetic sex pheromone	Higher parasitization, attraction	Abdelgader and Mazomenos (2002)
	Prays pleae	Sex pheromone components	Attraction in Y-tube olfactometer	(Milonas et al. 2003)
T. evanescens <u>later</u> *T. maidis* (= *T. brassicae*)	*Mamestra brassicae*	Calling virgin females Host sex pheromone odor No response to main sex pheromone component	Arrestment in 4-arm-olfactometer Arrestment in windtunnel	Noldus & van Lenteren (1985a); Noldus et al. (1991a)
	Pieris brassicae	Virgin females in 'mate-acceptance posture' no sex pheromone known	Arrestment in 4-arm olfactometer	Noldus & van Lenteren (1985a)
T. chilonis	*Helicoverpa assulta*	Sex pheromone of virgin moths Main component of sex pheromone	Arrestment in 4-arm olfactometer	Boo & Yang (2000)
	Plutella xylostella	Synthetic sex pheromone blend	Attraction in Y-tube olfactometer	Reddy et al. (2002)
T. cordubensis	*Heliothis armigera Earias insulana*	Adult female moth	Attraction in T-tube olfactometer	Noldus, (1989)
T. sp. p. buesi	*Ephestia kuehniella*	Adult female moth	Attraction in T-tube olfactometer	Noldus (1989)

Table 2: Phoretic egg-parasitoids

Egg parasitoid	Host	Body part attached	Transporting sex	Host cues	Host's reproduction	Reference
Scelionidae						
Synoditella spec.	Dichromorpha viridis	Wings or hind legs	Gravid females	Unknown	12-60 eggs	Clausen (1976)
S. fulgidus	Chortoicetes terminifera	Abdomen	Ovipositing females	Unknown	30-60 eggs	Clausen (1976)
S. bisulcus	Melanoplus spec.	Abdomen	Females	Melanonic scars	Egg pods	Clausen (1976)
S. viatrix	Colemania sphenarioides	unknown	Unknown	Unknown	Gregarious	Clausen (1976)
S. viatrix	Neorthacris acuticeps milgirensis	Abdomen	Females	Unknown	???	Clausen (1976)
Lepidoscelio spec.	N.simulans	Abdomen	Unknown	Unknown	???	Clausen (1976)
Matibaria manticida	Mantis religiosa	Wings/genital plates	Both sexes	Unknown	100-300 eggs	Couturier (1941)
Telenomus beneficiens	Schoenobius incertulus	Wings	Females	Unknown	Gregarious	Clausen (1976)
T. gracilis	Dendrolimus sibiricus	Thorax	Mated females	Unknown	Gregarious	Clausen (1976)
T. dendrolimusi	D. sibiricus albolineatus	Wings	unknown	Unknown	Gregarious	Tabata and Tamanuki (1940)
T. euproctidis	Euproctis taiwana E. pseudoconspersa	Anal tuft	Virgin females	Moth's sex pheromone	188.3 ± 24 eggs	Arakaki (1990); Arakaki et al. (1995); (1996)
T. calvus	Podisus maculiventris	Pronota and femora	Both sexes	Male pheromone, specific volatiles of mated females	17-70 eggs	Bruni et al. (2000); Orr et al. (1986); Aldrich et al. (1984), Buschman & Whitcomb (1980)
Telenomus spec.	Apantesis parthenice	Wings	Females	Unknown	Gregarious	Platt (1985)
Epinomus anoplocnemidis	Anoplocnemus curvipes	Head	unknown	Unknown	Gregarious	Clausen (1976)
Asolcus sulmo	Cantheconidea gaugleri	Head, antennae, thorax, legs	unknown	Unknown	???	Clausen (1976)
Boyeria vinosa	Calotelea n. sp. Boyeria	Thorax	Females	Unknown	???	Clausen (1976)

Protelenemus sp.	Anoplocnemis phasiana	Hind femora in males Antennae in females	Both sexes	Unknown	Eggs in batches	Kohno (2002)
Trichogrammatidae						
Oligosita xiphidii	Conocephalus longipenne	Hind wings	Females	Unknown	???	Clausen (1940)
O. brevicilia	Neoconocephalus sp.	Hind wings	unknown	Unknown	???	De Santis & Cicchino (1992)
Pseudoxer ufens forsysthi (=Xenufens sp.)	Caligo eurilochus	Hind wings	Both sexes	Unknown	Small batches	Malo (1961)
P. forsysthi	Opsiphanes cassina	Abdomen	unknown	Unknown	In clusters	Yoshimoto (1976)
Brachista efferiae B. fisheri	Efferia efferiae E. fisheri	Near base of halteres	Females	Unknown	???	Pinto (1994)
Trichogramma dendrolimi	D. sibiricus albolineatus	wings	unknown	Unknown	Gregarious	Tabata & Tamanuki (1940)
Trichogramma brassicae	Pieris brassicae	wings	Mated females	Antiphrodisiac	Gregarious	Fatouros et al. (2005b)
Ecyrtidae						
Taftia prodeniae	Dolichoderus bituberculatus	Bases of antennae	unknown	Unknown	???	Clausen (1976)
Eupelmidae						
Anastatus bifasciatus	Porthetria dispar	Abdomen	Females	Unknown	100-1000 eggs	Clausen (1976)
Eulophidae						
Emersonella riveipes	Stoias sp. Banotochara pavonia	Elytrae	Ovipositing females	Unknown	Stoias: 10-50	Carroll (1978)
Chrysocharodes rotundiventris	Mecistomela marginata	Unknown	Mated females	Unknown	Single/Gregarious	Macedo et al. (1990)
Grassator viater	Naupactus xanthographus	Prothorax	unknown	Unknown	Egg clusters	Clausen (1976)
Torymidae						
Podagrion fraternum	Mantis prasina Polyspilota pustulata	Hind wings	Females	Unknown	Egg masses	Clausen (1976)

3.1   Dietary specialization in egg parasitoids

Egg parasitoids like *Trichogramma* species are known to be fairly polyphagous, parasitizing a broad range of hosts specifically of the insect order Lepidoptera (Pinto and Stouthamer 1994; Thomson and Stinner 1990). This could be considered as one solution to solve the major constraints that are set by their life history (Vet et al. 1995). However, host ranges of *Trichogramma* spp. are mainly determined in the laboratory, where they tend to be broader than in the field (reviewed by Romeis et al. 2005). Still, there is growing evidence for *Trichogramma* spp. being more prevalent in certain habitats or on specific plants (reviewed by Romeis et al. 2005) probably due to their limited moving abilities. Plant structure, host density and distribution, and other (micro-) habitat specificities might vary and influence the host foraging behavior of egg parasitoids (Gingras and Boivin 2002; Romeis et al. 2005). This habitat/ plant loyalty could narrow their host range. Certainly, egg parasitoids in monocultures or habitats of low biodiversity would encounter only a small range of host species. The uncertainty of the host range in the field, as well as wrong species determinations in the past makes it difficult to put *Trichogramma* spp. into categories as generalists or specialists. This means that the dietary specialization aspects of Vet and Dicke (1992) are difficult to apply on *Trichogramma* spp. as long as their host specificity remains poorly known.

Here, we try to find a general concept for egg parasitoids, dealing with the major constraints of egg parasitoids concerning the host like: a small size, sessile and inconspicuous life, and short-time availability. Next to a generally short longevity, egg parasitoids, like *Trichogramma* spp. are known to be restricted to passive downwind movements or walking and jumping. Which are the main infochemical strategies used in their long-distance search for hosts?

3.2   The three major strategies used by egg parasitoids

3.2.1   *The ability to learn infochemical cues*

According to Vet et al. (1990) a response to infochemicals was considered as learned when it was observed in only experienced insects after having encounters with host/prey or host plant/feeding substrates. Associative learning by connecting an unconditioned stimulus, always host derived, with a conditioned stimulus is frequently observed in parasitoids. Generally, responses of carnivores to host-derived kairomones are fixed, whereas responses to plant volatiles are more variable and are learned mostly during adult foraging (Vet and Dicke 1992). Immature parasitoids can also gain experience through development inside the host (Corbet 1985).

Vet et al. (1995) stated that learning of foraging cues by egg parasitoids is not expected. The more foraging decisions are done the more likely infochemical learning becomes adaptive. The high availability of eggs (compared to other stages such as larvae or pupae) and the short longevity of egg parasitoids in combination with their limited ability to search may restrain the number of foraging decisions that lead to host encounters. They hypothesized that if learning occurs then this is most likely to occur in the more polyphagous egg parasitoid species that attack hosts that lay single eggs.

However, several recent studies show the importance of learning of plant cues during foraging in egg parasitoids. For example, experience during emergence modified the searching behavior of one of the two tested *T.* nr *brassicae* strains towards plant stimuli: females that emerged from their host on a tomato plant subsequently searched a tomato leaf longer than females emerging in a vial without plant stimuli (Bjorksten and Hoffmann 1998). When comparing a pre-adult experience with the effect of adult learning through an oviposition experience on host preference in two strains of *T.* nr *ivelae*, it was shown that adult experience had a stronger effect than learning through development inside the host (Bjorksten and Hoffmann 1995).

Oviposition-induced plant cues need to be learned by some egg parasitoid species, while others are able to respond innately. For example, the generalist *T. brassicae* needs an oviposition experience with host eggs on Brussels spout leaves before it responds to the plant synomone induced by the eggs of *P. brassicae* (Fatouros et al. 2005a) (Chapter 4). That same implies for *C. ruforum* which needs to learn plant synomones of sawfly egg-laden pine trees (Mumm et al. 2005). In contrast, neither *Oomyzus gallerucae* nor *Trissolcus basalis* need to learn the oviposition-induced plant synomones (see 1.1.2). While *O. gallerucae* is a specialized egg parasitoid, *T. basalis* attacks eggs of several phytophagous and predatory pentatomid bugs such as *Nezara viridula*, a highly polyphagous host. Despite this polyphagy at the host and host plant level, *Tr. basalis* uses the oviposition-induced synomones of the bean plant innately to locate eggs of *N. viridula* (Colazza et al. 2004). The results on the tritrophic systems studied so far with respect to oviposition-induced plant cues indicate that egg parasitoids need to learn these plant synomones when they only quantitatively differ from non-induced plants, while they are able to respond innately when there is a qualitative change of plant volatiles induced by egg deposition. Further studies are needed to elucidate whether this is a general pattern also for other systems.

In contrast to plant cues, egg parasitoids were generally found to respond innately to host cues. For example, *Trichogramma brassicae* preferred *P. brassicae* females treated with the synthetic anti-sex pheromone benzyl cyanide over untreated females innately (Fatouros et al. 2005b) (Chapter 3). The same implies for the specialist *Chrysonotomyia ruforum*: it detects the sawfly sex pheromone without having to learn the odor beforehand (Hilker et al. 2000). But even for host cues it has been shown that some egg parasitoids have to learn them associatively. Females of *T. evanescens* responded to the synthetic main component of pyralid moths sex pheromone (Z,E)-9,12-tetradecadenyl acetate (ZETA) only when they had an oviposition experience in vials with ZETA-loaded air (Schöller and Prozell 2002). Kaiser et al. (1989) tested the orientation responses of *Trichogramma maidis* (=*T. brassicae*) to airborne odors such as host cues and plant extracts offered singly or in combination to naïve wasps and oviposition-experienced wasps exposed to tested odor. Inexperienced females failed to respond to the synthetic pheromone of the host *Ostrinia nubilalis*, a mixture of (Z)-11 and (E)-11-tetradecenyl-acetate. However, naïve conspecific wasps did react to a mixture of egg odors, the synthetic pheromone, and a maize plant extract. An oviposition experience together with an exposure to the maize odor or a combination of odors increased the response to the conditioning scent. It was suggested that *T. maidis* can learn to associate some olfactory cues with the presence of the host. Other reports on *Trichogramma* spp. even show that an experience has no effect on the response to the pheromone (McGregor and Henderson 1998).

3.2.2 The importance of the adult host stage

Egg parasitoids seem to heavily rely on the adult host stage: sex pheromones of the host female have a large impact on indirectly locating eggs of moths, butterflies or sawflies (Fatouros et al. 2005b; Hilker et al. 2000; Noldus 1989; Powell 1999). This infochemical detour is an important strategy used by many egg parasitoids. Only a few larval or pupal parasitoids were reported to respond to sex pheromones. Some adult or larval parasitoids may use sex pheromones of homopterans or aggregation pheromones of adult bark beetles, heteropteran bugs or fruit flies (reviewed by Powell 1999).

Besides an arresting effect to areas likely to contain host eggs, (anti-)sex pheromones may have a much greater importance to phoretic egg parasitoids.

Phoresy is exclusively performed by egg parasitoids, which are smaller when compared to other parasitoids and therefore often limited in their flight capabilities. From the literature available on infochemical use in egg parasitoids, chemical espionage either or not in combination with phoresy seems to be the most commonly used solution of egg parasitoids to the reliability-detectability problem.

3.2.3 Plant volatiles induced by egg deposition or feeding

Egg parasitoids developed to use a third so far underestimated solution, mentioned by Vet and Dicke (1992): plant volatiles induced by herbivore infestation. This could solve the detectability problem that egg parasitoids face. Here, they are known to follow two different strategies: plant volatiles induced by herbivore feeding or by egg deposition. Both types of plant cues can be very specific in indicating the presence of host eggs, and thus might play equally important roles for the host location process in egg parasitoids.

From the plant's point of view, an attraction of parasitoids/predators to eggs deposited on the plant might have a great advantage because hatching larvae can be killed before they start feeding. Especially annual plants (with a smaller biomass and shorter life time) are expected to benefit from such an indirect defense mechanism (Hilker and Meiners 2002b). From the egg parasitoid´s perspective it has remained unclear so far when it is beneficial to use oviposition-induced plant cues to locate their hosts, and when the use of feeding-induced plant volatiles becomes adaptive. Generally, plant volatiles induced by feeding damage should be more abundant due to the greater plant damage than plant volatiles induced after egg deposition by herbivores. However, a response to feeding-induced plant volatiles should be only adaptive to egg parasitoids when they provide reliable information, e.g., when the host generations overlap and all co-occur on the plant. On the other hand, a response of egg parasitoids to oviposition-induced plant volatiles requires fine-tuned perceptive abilities, most likely present mainly in systems where tritrophic levels specialized on each other over evolutionary time.

We expect such an indirect plant response towards egg deposition in those systems where a) wounding is involved during oviposition, b) the egg parasitoid has flight capabilities, and c) stimuli are presented in an olfactory background. If no simultaneous wounding occurs (either by egg deposition or by feeding of adults or larvae), we expect no induction of volatiles by the plant but short-range cues, likely to be restricted to the plant surface and perceived by the egg parasitoid only in close vicinity (see Fatouros et al. 2005a) (Chapter 4). In this case other cues than plant volatiles need to act as long-range attractants, such as anti-aphrodisiacs guiding the egg parasitoid to the transporting host and future host eggs (compare Fatouros et al. 2005b) (Chapter 3).

In studies on the response of *C. ruforum* to oviposition-induced pine odors it was shown that the parasitoids were only attracted to the main compound of the plant volatile bouquet, (*E*)-β-farnesene when offered in a mixture with odors from egg-free pines. The chemical itself was not attractive at any concentration tested (Mumm and Hilker 2005; Mumm et al. 2003). In contrast, the egg parasitoid of the elm leaf beetle, *O. gallerucae*, responds to the oviposition-induced compound 4,8-dimethyl-1,3,7-nonatriene (DMNT) alone both in the laboratory and in the field. Here, an intense background odor of mainly green leaf volatiles might hamper the orientation towards the synomone (Meiners, unpublished results). The induction of plant volatiles by oviposition involves rather small changes in the plant odor bouquet. Within the complex of plant volatiles minute modifications of single compounds in a background context might be sufficient for an egg parasitoid to detect an herbivore-infested host plant. However, more studies are needed to examine this hypothesis.

3.3 General patterns in egg parasitoid foraging strategies

To reach (new) habitats egg parasitoids have to move over larger distances. They can achieve this by spreading passively by wind or by using larger hosts for transportation (phoresy) (Figure 1). Both ways implicate a momentum of chance either by coming down at the right spot in the landscape or sitting on the right place to catch a suitable host for transportation. However, egg parasitoids might use general abiotic or biotic habitat cues or cues of host encounter places (like flower heads) to enhance the chance to reach a suitable habitat.

We expect directed orientation of egg parasitoids over a smaller distance. In closer vicinity to host eggs, both host cues (in large enough quantities) and plant cues might play equally important roles for the host location process depending on factors such as learning capabilities or the dietary specialization of the egg parasitoid and its host.

To fully understand infochemical exploiting strategies used by foraging egg parasitoids it is necessary to study plant-host-parasitoid interactions at different scales in the field. After having detected the scales that are important for host location, the mechanism of infochemical host search should be investigated experimentally. Furthermore, behavioral and electrophysiological investigations should provide information on the chemicals egg parasitoids can use.

Conclusions

Egg parasitoids play an important role in biological control programs of pest insects. Understanding their host location behavior is a crucial step for a targeted application of parasitoids in crop fields and greenhouses. This paper summarizes the extensive research so far done on the role of infochemicals used by egg parasitoids during their host foraging behavior. However, despite of the

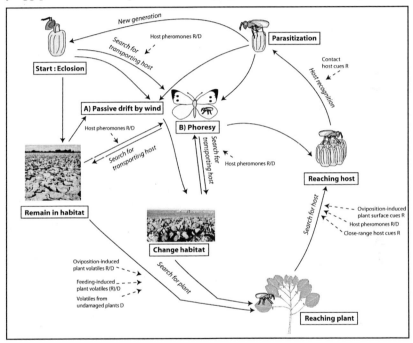

Figure 1: The general host location behavior of egg parasitoids and the possible infochemicals involved are shown on the basis of a virtual system. After eclosion an egg parasitoid can choose to remain in the same habitat and search for a host plant or to change to a new habitat. To reach habitats the wasp has two possibilities to spread: a) by chance and drift passively with the wind or b) by searching adult hosts for transportation (phoresy). By using cues of the adult host, either to directly find the host plant and deposited herbivorous eggs or to achieve phoresy, egg parasitoids are shown to rely on the adult host stage more than other parasitoids do. Besides host pheromones, herbivore-induced plant cues (egg deposition or feeding) seem to have a so far underestimated impact on the host foraging behavior of egg parasitoids (see text for more details). Dashed arrows indicate a signal used by the egg parasitoid. R = reliable cue, D = detectable cue.

many publications on this topic, there are still important gaps in our knowledge, especially concerning the behavior of these tiny wasps in nature. A general pattern of strategies could be highlighted in this paper: due to the inconspicuousness of host eggs, egg parasitoids have developed so-called infochemical detour strategies to locate their hidden hosts. This infochemical detour via specific cues from the adult host stage, sometimes in combination with phoresy, or via cues from plants induced by larval or adult feeding implies a widespread use of very host specific infochemicals within egg parasitoids. The use of oviposition-induced plant cues indicating the presence of the host eggs is another elegant solution to the reliability-detectability problem. A previous experience with the plant stimuli in association with a host cue seems to play an important role in the exploitation of induced plant cues. The innate response to specific host cues of polyphagous parasitoids is seen to be unlikely because of physiological constraints caused by the diversity of specific cues expected from several hosts (Vet and Dicke, 1992). Yet, it cannot be excluded that these constraints are less severe than expected and that generalists with a smaller host range are indeed able to innately respond to the specific information from different hosts (Steidle and van Loon 2002). However, an uncertainty of the host range in the field makes it difficult to put *Trichogramma* spp. into categories as generalists or specialists and thus apply the concept of dietary specialization of Vet and Dicke (1992) on egg parasitoids as long as their host specificity remains uncertain.

References

Abdelgader H, Mazomenos B (2002) Response of *Trichogramma oleae* (Hymenoptera: Trichogrammatidae), to host pheromones, frass and scales extracts. Egg Parasitoid News 14:16-17

Aldrich JR (1985) Pheromone of a true bug (Hemiptera-Heteroptera): Attractant for the predator, *Podisus maculiventris*, and kairomonal effects. In: Acree TE, Sonderlund DM (eds) Semiochemistry: Flavors and Pheromones. de Gruyter, Berlin, pp 95-119

Aldrich JR (1995) Chemical communication in the true bugs and parasitoid exploitation. In: Cardé RT, Bell WJ (eds) Chemical Ecology of Insects II. Chapman and Hall, New York, pp 318-363

Aldrich JR, Kochansky JP, Abrams CB (1984) Attractant for beneficial insect and its parasitoids: pheromone of the predatory spined soldier bug, *Podisus maculiventris* (Hemiptera: Pentatomidae). Environmental Entomology 13:1031-1036

Altieri MA, Annamalai S, Katiyar KP, Flath RA (1982) Effects of plant extracts on the rates of parasitization of *Anagasta kuehniella* (Lep.: Pyralidae) eggs by *Trichogramma pretiosum* (Hym.: Trichogrammatidae) under greenhouse conditions. Entomophaga 27:431-438

Andersson J, Borg-Karlson A-K, Wiklund C (2003) Antiaphrodisiacs in Pierid butterflies: a theme with variation! Journal of Chemical Ecology 29:1489-1499

Andersson J, Borg-Karlson A-K, Wiklund C (2004) Sexual conflict and anti-aphrodisiac titre in a polyandrous butterfly: male ejaculate tailoring and absence of female control. Proceedings of the Royal Society: Biological sciences 271:1765-1770

Arakaki N, Wakamura S (2000) Bridge in time and space for an egg parasitoid-kairomonal use of trace amount of sex pheromone adsorbed on egg mass scale hair of the tussock moth, *Euproctis taiwana* (Shiraki) (Lepidotera: Lymantriidae), by an egg parasitoid, *Telenomus euproctidis* Wilcox (Hymenoptera: Scelionidae), for host location. Entomological Science 3: 25-31

Arakaki N, Wakamura S, Yasuda T (1995) Phoresy by an egg parasitoid, *Telenomus euproctidis* (Hymenoptera: Scelionidae), on the tea tussock moth, *Euproctis pseudoconspersa* (Lepidoptera: Lymantriidae). Applied Entomology and Zoology 30:602-603

Arakaki N, Wakamura S, Yasuda T (1996) Phoretic egg parasitoid, *Telenomus euproctidis* (Hymenoptera: Scelionidae), uses sex pheromone of tussock moth *Euproctis taiwana* (Lepidoptera: Lymantriidae) as a kairomone. Journal of Chemical Ecology 22:1079-1085

Arakaki N, Wakamura S, Yasuda T, Yamagishi K (1997) Two regional strains of a phoretic egg parasitoid, *Telenomus euproctidis* (Hymenoptera: Scelionidae), that use different sex pheromones of two allopatric tussock moth species as kairomones. Journal of Chemical Ecology 23:153-161

Bai SX, Wang ZY, He KL, Zhou DR (2004) Olfactory responses of *Trichogramma ostriniae* Pang et. Chen to kairomones from eggs and different stages of adult females of *Ostrinia furnacalis* (Guenee). Acta Entomologica Sinica 47:48-54

Bar D, Gerling D, Rossler Y (1979) Bionomics of the principal natural enemies attacking *Heliothis armigera* in cotton fields in Israel. Enviromental Entomolgy 8:468-474

Battisti A (1989) Field studies on the behaviour of two egg parasitoids of the pine processionary moth *Thaumetopoea pityocampa*. Entomophaga 34:29-38

Begum M, Gurr GM, Wratten SD, Nicol HI (2004) Flower color affects tri-trophic-level biocontrol interactions. Biological Control 30:584-590

Bezzerides A et al. (2004) Plant-derived pyrrolizidine alkaloid protects eggs of a moth (*Utetheisa ornatrix*) against a parasitoid wasp (*Trichogramma ostriniae*). Proceedings of the National Academy of Sciences of the United States of America 101:9029-9032

Bin F, Vinson SB, Strand MR, Colazza S, Jones Jr. WA (1993) Source of an egg kairomone for *Trissolcus basalis*, a parasitoid of *Nezara viridula*. Physiological Entomology 18:7-15

Bjorksten TA, Hoffmann AA (1995) Effects of pre-adult and adult experience on host acceptance in choice and non-choice tests in two strains of *Trichogramma*. Entomologia Experimentalis et Applicata 76:49-58

Bjorksten TA, Hoffmann AA (1998) Plant cues influencing searching behaviour and parasitism in the egg parasitoid *Trichogramma* nr. *brassicae*. Ecological Entomology 23:355-362

Blaakmeer A, Hagenbeek D, van Beek TA, de Groot AE, Schoonhoven LM, van Loon JJA (1994a) Plant response to eggs vs. host marking pheromone as factors inhibiting oviposition by *Pieris brassicae*. Journal of Chemical Ecology 20:1657-1665

Blaakmeer A et al. (1994b) Isolation, identification, and synthesis of miriamides, new hostmarkers from eggs of *Pieris brassicae*. Journal of Natural Products 57:90-99

Blum MS, Hilker M (2002) Chemical protection of insect eggs. In: Hilker M, Meiners T (eds) Chemoecology of Insect Eggs and Egg Deposition. Blackwell Publishing Ldt, Oxford, pp 61-90

Boo KS, Yang JP (1998) Olfactory response of *Trichogramma chilonis* to *Capsicum annum*. Journal of Asia-Pacific Entomology 1:123-129

Boo KS, Yang JP (2000) Kairomones used by *Trichogramma chilonis* to find *Helicoverpa assulta* eggs. Journal of Chemical Ecology 26:359-375

Bruni R, Sant'Ana J, Aldrich JR, Bin F (2000) Influence of host pheromone on egg parasitism by scelionid wasps: Comparison of phoretic and nonphoretic parasitoids. Journal of Insect Behavior 12:165-173

Buschman LL, Whitcomb WH (1980) Parasites of *Nezara viridula* (Hemiptera: Pentatomidae) and other Hemiptera in Florida. Florida Entomologist 63:154-162

Carroll RC (1978) Beetles, parasitoids and tropical morning glories: a study in host discrimination. Ecological Entomology 3:79-85

Clausen CP (1940) Entomophagous Insects. McGraw-Hill, New York

Clausen CP (1976) Phoresy among entomophagous insects. Annual Review of Entomology 21:343-368

Colazza S, Fucarino A, Peri E, Salerno G, Conti E, Bin F (2004a) Insect oviposition induces volatile emission in herbaceous plants that attracts egg parasitoids. Journal of Experimental Biology 207:47-53

Colazza S, McElfresh JS, Millar JG (2004b) Identification of volatile synomones, induced by *Nezara viridula* feeding and oviposition on bean spp., that attract the egg parasitoid *Trissolcus basalis*. Journal of Chemical Ecology 30:945-964

Colazza S, Rosi CM, Clemente A (1997) Response of egg parasitoid *Telenomus busseolae* to sex pheromone of *Sesamia nonagrioides*. Journal of Chemical Ecology 23:2437-2444

Colazza S, Salerno G, Wajnberg E (1999) Volatile and contact chemicals released by *Nezara viridula* (Heteroptera:Pentatomidae) have a kairomonal effect on the egg parasitoid *Trissolcus basalis* (Hymenoptera:Scelionidae). Biological Control 16:310-317

Conti E, Salerno G, Bin F, Williams HJ, Vinson SB (2003) Chemical cues from *Murgantia histrionica* eliciting host location and recognition in the egg parasitoid *Trissolcus brochymenae*. Journal of Chemical Ecology 29:115-130

Corbet SA (1985) Insect chemosensory responses: a chemical legacy hypothesis. Ecological Entomology 10:143-153

Couturier A (1941) Nouvelles observations sur *Rielia manticida* Kief., hyménoptère parasite de la mante religieuse. II. Comportement de l'insecte parfaite. Revue de Zoologie Agricole et Appliquee 40:49-62

De Santis L, Cicchino C (1992) Foresis por *Oligosita brevicilia* (Hymenoptera) sobre *Neoconocephalus* sp. (Orthopthera) en la República Argentina. Anales de la Academia Nacional de Agronomia y Veterinaria 46:5-7

Dicke M, Sabelis MW, Takabayashi J, Bruin J, Posthumus MA (1990) Plant strategies of manipulating predator-prey interactions through allelochemicals: prospects for application in pest control. Journal of Chemical Ecology 16:3091-3118

Dicke M, van Loon JJA (2000) Multitrophic effects of herbivore-induced plant volatiles in an evolutionary context. Entomologia Experimentalis et Applicata 97:237-249

Dicke M, Vet LEM (1999) Plant-carnivore interactions: evolutionary and ecological consequences for plant, herbivore and carnivore. In: Olff H, Brown VK, Drent RH (eds) Herbivores: Between Plants and Predators. Blackwell Science, Oxford, pp 483-520

Fatouros NE, Bukovinszkine'Kiss G, Kalkers LA, Soler Gamborena R, Dicke M, Hilker M (2005a) Oviposition-induced plant cues: do they arrest *Trichogramma* wasps during host location? Entomologia Experimentalis et Applicata 115:207-215

Fatouros NE, Huigens ME, van Loon JJA, Dicke M, Hilker M (2005b) Chemical communication - Butterfly anti-aphrodisiac lures parasitic wasps. Nature 433:704

Fiaboe MK, Chabi-Olaye A, Gounou S, Smith H, Borgemeister C, Schultheiss F (2003) *Sesamia calamistis* calling behavior and its role in host finding of egg parasitoids *Telenomus busseolae*, *Telenomus isis*, and *Lathromeris ovicida*. Journal of Chemical Ecology 29:921-929

Fournier F, Boivin G (2000) Comparative dispersal of *Trichogramma evanescens* and *Trichogramma pretiosum* (Hymenoptera: Trichogrammatidae) in relation to enviromental conditions. Environmental Entomology 29:55-63

Frenoy C, Durier C, Hawlitzky N (1992) Effect of kairomones from egg and female adult stages of *Ostrinia nubilalis* (Hübner) (Lepidoptera, Pyralidae) on *Trichogramma brassicae* Bezdenko (Hymenoptera, Trichogrammatidae) female kinesis. Journal of Chemical Ecology 18:761-

773

Garnier-Geoffroy F, Robert P, Hawlitzky N, Frerot B (1996) Oviposition behaviour in *Ostrinia nubilalis* (Lep.: Pyralidae) and consequences on host location and oviposition in *Trichogramma brassicae* (Hym.: Trichogrammatidae). Entomophaga 41:287-299

Gillott C (2002) Insect accessory reproductive glands: key players in production and protection of eggs. In: Hilker M, Meiners T (eds) Chemoecology of Insect Eggs and Egg Deposition. Blackwell Publishing LTd, Oxford, pp 37-59

Gingras D, Boivin G (2002) Effect of plant structure, host density and foraging duration on host finding by *Trichogramma evanescens* (Hymenoptera: Trichogrammatidae). Environmental Entomology 31:1153-1157

Godfray HCJ (1994) Parasitoids - Behavioral and Evolutionary Ecology. Princeton University Press, Princeton, New Jersey

Greenfield MD (1981) Moth sex pheromones: an evolutionary perspective. Florida Entomologist 64: 4-17

Gross JR. HR, Lewis WJ, Beevers M, Nordlund DA (1984) *Trichogramma pretiosum* (Hymenoptera: Trichogrammatidae): Effects of augmented densities and distributions of *Heliothis zea* (Lepidoptera: Noctuidae) host eggs and kairomones on field performance. Envirom.Entomol. 13:981-985

Hilker M (1994) Egg deposition and protection of eggs in Chrysomelidae. In: Jolivet PH, Cox ML, Petitpierre E (eds) Novel Aspects of the Biology of Chrysomelidae. Kluwer Academic Publishers, Amsterdam, pp 263-276

Hilker M, Bläske V, Kobs C, Dippel C (2000) Kairomonal effects of the sawfly sex pheromones of egg parasitoids. Journal of Chemical Ecology 26:2591-2601

Hilker M, Kobs C, Varama M, Schrank K (2002) Insect egg deposition induces *Pinus sylvestris* to attract egg parasitoids. Journal of Experimental Biology 205:455-461

Hilker M, Meiners T (1999) Chemical Cues Mediating Interactions Between Chrysomelids and Parasitoids. In: Cox ML (ed) Advances in Chrysomelidae Biology, vol 1. Backhuys Publishers, Leiden, The Netherlands, pp 197-216

Hilker M, Meiners T (eds) (2002a) Chemoecology of Insect Eggs and Egg Deposition. Blackwell, Berlin

Hilker M, Meiners T (2002b) Induction of plant responses towards oviposition and feeding of herbivorous arthropods: a comparison. Entomologia Experimentalis et Applicata 104:181-192

Hilker M, Meiners T (2006) Early herbivore alert: insect eggs induce plant defense. Journal of Chemical Ecology:in press

Hilker M, Stein C, Schröder R, Varama M, Mumm R (2005) Insect egg deposition induces defence responses in *Pinus sylvestris*: characterisation of the elicitor. Journal of Experimental Biology 208:1849-1854

Hinton HE (1981) Biology of Insect Eggs. Pergamon Press, Oxford

Howard LO (1927) Concerning phoresy in insects. Entomological News 38:145-147

Jones RL, Lewis WJ, Beroza M, Bierl BA, Sparks AN (1973) Host-seeking stimulants (kairomones) for the egg parasite, *Trichogramma evanescens*. Environmental Entomology 2:593-596

Kaiser L, Pham-Delegue MH, Bakchine E, Masson C (1989) Olfactory responses of *Trichogramma maidis* Pint. et Voeg.: Effects of chemical cues and behavioral plasticity. Journal of Insect Behavior 2:701-712

Keller MA, Lewis WJ, Stinner RE (1985) Biological and practical significance of movement by *Trichogramma* species: a review. Southwestern Entomologist 8:138-155

Kennedy BH (1984) Effect of multilure and its components on parasites of *Scolytus multistriatus* (Coleoptera: Scolytidae). Journal of Chemical Ecology 10:373-385

Klemm U, Schmutterer H (1993) Wirkungen von Niempräparaten auf die Kohlmotte *Plutella xylostella* L. und ihre natürlichen Feinde der Gattung *Trichogramma*. Zeitschrift für Pflanzenkrankheiten und Pflanzenschutz 100:113-128

Klijnstra JW (1986) The effect of an oviposition deterring pheromone on egglaying in *Pieris brassicae*. Entomologia Experimentalis et Applicata 41:139-146

Kohno K (2002) Phoresy by an egg parasitoid, *Protelenomus sp.* (Hymenotpera: Scelionidae), on the coreid bug *Anoplocnemis phasiana* (Heteroptera: Coreidae). Entomological Science 5:281-285

Laing J (1937) Host-finding by insect parasites. 1. Observations on the finding of hosts by *Alysia manducator, Mormoniella vitripennis* and *Trichogramma evanescens*. Journal of Animal Ecology 6:298-317

Leal WS, Higushi H, Mizutani N, Nakamori H, Kadosawa T, Ono M (1995) Multifunctional communication in *Riptortus clavatus* (Heteroptera: Alydidae): conspecific nymphs and egg parasitoid *Ooencyrtus nezarae* use the same adult attractant pheromone as chemical cue. Journal of Chemical Ecology 211:973-985

Leonard DE, Wu Z-X, Ferro D, N. (1987) Responses of parasite *Edovum puttleri* to kairomone from eggs of Colorado potato beetle, *Leptinotarsa decemlineata*. Journal of Chemical Ecology 13: 325-344

Lewis WJ, Jones RL, Nordlund DA, Gross JR. HR (1975) Kairomones and their use for mangament of entomophagous insects. II. Mechanisms causing increase in rate of parasitization by *Trichogramma* spp. Journal of Chemical Ecology 1:349-360

Lewis WJ, Jones RL, Sparks AN (1972) A host-seeking stimulant for the egg parasite, *Trichogramma evanescens*: its source and a demonstration of its laboratory and field activity. Annals of the Entomological Society of America 65:1087-1089

Lewis WJ, Nordlund DA, Gueldner RC, Teal PEA, Tumlinson JH (1982) Kairomones and their use for management of entomophagous insects. XIII. Kairomonal activity for *Trichogramma* spp. of abdominal tips, excretion, and a synthetic sex pheromone blend of *Heliothis zea* (Boddie) moths. Journal of Chemical Ecology 8:1323-1331

Lewis WJ, Sparks AN, Redlinger LM (1971) Moth odor: A method of host-finding by *Trichogramma evanescens*. Journal Economic Entomology 64:557-558

Li LY (1994) Worldwide use of *Trichogramma* for biological control on different crops: A survey. In: Wajnberg E, Hassan SA (eds) Biological Control with Egg Parasitoids. CAB International, Wallingford, pp 37-53

Lou YG, Ma B, Cheng JA (2005) Attraction of the parasitoid *Anagrus nilaparvatae* to rice volatiles induced by the rice brown planthopper *Nilaparvata lugens*. Journal of Chemical Ecology 31: 2357-2372

Macedo MV, de Santis L, Monteiro RF (1990) *Chrysocharodes rotundiventris* De Santis, new species (Eulophidae), a phoretic parasitoid, with notes on its ecology and behaviour. Revista Brasileira de Entomologia 34:637-641

Malo F (1961) Phoresy in *Xenufens* (Hymenoptera: Trichogrammatidae), a parasite of *Caligo eurilochus* (Lepidoptera: Nymphalidae). Journal of Economic Entomology 54:465-466

Mattiacci L, Vinson SB, Williams HJ (1993) A long-range attractant kairomone for egg parasitoid *Trissolcus basalis*, isolated from defensive secretion of its host, *Nezara viridula*. Journal of Chemical Ecology 19:1167-1181

McGregor R, Henderson D (1998) The influence of oviposition experience on response to host pheromone in Trichogramma sibericum (Hymenoptera: Trichogrammatidae). Journal of Insect Behavior 11:621-632

Meiners T, Hilker M (1997) Host location in *Oomyzus gallerucae* (Hymenoptera: Eulophidae), an egg parasitoid of the elm leaf beetle *Xanthogaleruca luteola* (Coleoptera: Chrysomelidae). Oecologia 112:87-93

Meiners T, Hilker M (2000) Induction of plant synomones by oviposition of a phytophagous insect. Journal of Chemical Ecology 26:221-232

Meiners T, Köpf A, Stein C, Hilker M (1997) Chemical signals mediating interactions between Galeruca tanaceti L. (Coleoptera, Chrysomelidae) and its egg parasitoid Oomyzus galerucivorus (Hedqvits) (Hymenoptera, Eulophidae). Journal of Insect Behavior 10:523-539

Milonas P, Konstantopoulou M, Mazomenos B (2003) Kairomonal effect of sex pheromones and plant volatiles to *Trichogramma oleae*. Egg Parasitoid News 15:18-19

Moraes MCB, Laumann R, Sujii ER, Pires C, Borges M (2005) Induced volatiles in soybean and pigeon pea plants artificially infested with the neotropical brown stink bug, Euschistus heros, and their effect on the egg parasitoid, Telenomus podisi. Entomologia Experimentalis et Applicata 115:227-237

Mumm R, Hilker M (2005) The significance of background odour for an egg parasitoid to detect plants with host eggs. Chemical Senses 30:337-343

Mumm R, Schrank K, Wegener R, Schulz S, Hilker M (2003) Chemical analysis of volatiles emitted by *Pinus sylvestris* after induction by insect oviposition. Journal of Chemical Ecology 29: 1235-1252

Mumm R, Tiemann T, Varama M, Hilker M (2005) Choosy egg parasitoids: Specificity of oviposition-induced pine volatiles exploited by an egg parasitoid of pine sawflies. Entomologia Experimentalis et Applicata 115:217-225

Murlis J, Elkinton JS, Carde R (1992) Odor plumes and how insects use them. Annual Review of Entomology 37:305-332

Noldus LPJJ (1988) Response of the egg parasitoid *Trichogramma pretiosum* to the sex pheromone of its host *Heliothis zea*. Entomologia Experimentalis et Applicata 48:293-300

Noldus LPJJ (1989) Semiochemicals, foraging behaviour and quality of entomophagous insects for biological control. Journal of Applied Entomology 108:425-451

Noldus LPJJ, Lewis WJ, Tumlinson JH (1990) Beneficial arthropode behavior mediated by airborne semiochemicals. IX. Differential response of *Trichogramma pretiosum*, an egg parasitoid of *Heliothis zea*, to various olfactory cues. Journal of Chemical Ecology 16:3531-3544

Noldus LPJJ, Potting RPJ, Barendregt HE (1991a) Moth sex pheromone adsorption to leaf surface: bridge in time for chemical spies. Physiological Entomology 16:329-344

Noldus LPJJ, van Lenteren JC (1985a) Kairomones for the egg parasite *Trichogramma evanescens* Westwood I. Effect of volatile substances released by two of its hosts, *Pieris brassicae* L. and *Mamestra brassicae*, L. Journal of Chemical Ecology 11:781-791

Noldus LPJJ, van Lenteren JC (1985b) Kairomones for the egg parasite *Trichogramma evanescens* Westwood II. Effect of contact chemicals produced by two of its hosts, *Pieris brassicae* L. and *Pieris rapae* L. Journal of Chemical Ecology 11:793-800

Noldus LPJJ, van Lenteren JC, Lewis WJ (1991b) How *Trichogramma* parasitoids use moth sex pheromones as kairomones: orientation behaviour in a wind tunnel. Physiological Entomology 16:313-327

Nordlund DA (1994) Habitat location by *Trichogramma*. In: Wajnberg E, Hassan SA (eds) Biological Control with Egg Parasitoids. CAB International, Wallingford, Oxon, UK, pp 155-163

Nordlund DA, Chalfant RB, Lewis WJ (1985) Response of *Trichogramma pretiosum* females to volatile synomones from tomato plants. Journal of Entomological Science 20:372-376

Nordlund DA, Lewis WJ, Gueldner RC (1983) Kairomones and their use for management of entomophagous insects XIV. Response of *Telenomus remus* to abdominal tips of *Spodoptera frugiperda*, (Z)-9-tetradecene-1-ol acetate and (Z)-9-dodecene-1-ol acetate. Journal of Chemical Ecology 9:695-701

Nordlund DA, Strand MR, Lewis WJ, Vinson SB (1987) Role of kairomones from host accessory gland secretion in host recognition by *Telenomus remus* and *Trichogramma pretiosum*, with partial characterization. Entomologia Experimentalis et Applicata 44:37-44

Orr DB, Russin JS, Boethel DJ (1986) Reproductive biology and behavior of *Telenomus calvus* (Hymenoptera: Scelionidae), a phoretic egg parasitoid of *Podisus maculiventris* (Hemiptera: Pentatomidae). Canadian Entomologist 118:1063-1072

Padmavathi C, Paul AVN (1998) Saturated hydrocarbons as kairomonal source for the egg parasitoid, *Trichogramma chilonis* Ishii (Hym., Trichogrammatidae). Journal of Applied Entomology-Zeitschrift fuer Angewandte Entomologie 122:29-32

Paul AVN, Singh S, Singh AK (2002) Kairomonal effect of some saturated hydrocarbons on the egg parasitoids, *Trichogramma brasiliensis* (Ashmead) and *Trichogramma exiguum*, Pinto, Platner and Oatman (Hym., Trichogrammatidae). Journal Of Applied Entomology-Zeitschrift Fur Angewandte Entomologie 126:409-416

Pinto JD (1994) A taxonomic study of *Brachista* (Hymenoptera: Trichogrammatidae) with a description of two new species phoretic on robberflies of the genus *Efferia* (Diptera: Asilidae). Proceedings of the Entomological Society of Washington 96:120-132

Pinto JD, Stouthamer R (1994) Systematics of Trichogrammatidae with emphasis on *Trichogramma*. In: Wajnberg E, Hassan SA (eds) Biological Control with Egg parasitoids. CAB international, Oxon, pp 1-36

Platt AP (1985) Egg parasitsm of *Apantesis parthenice* (Arctiidae) through apparant phoresy by the wasp *Telenomus* sp. (Scelionidae). Journal of the Lepidopterists' Society 39:59-62

Powell W (1999) Parasitoid hosts. In: Hardie J, Minks AK (eds) Pheromones of Non-Lepidopteran Insects Associated with Agricultural Plants. CABI Publishing, Wallingford, pp 405-427

Quicke DLJ (1997) Parasitic Wasps. Chapman & Hall, London

Reddy GVP, Holopainen JK, Guerrero A (2002) Olfactory responses of *Plutella xylostella* natural enemies to host pheromone, larval frass, and green leaf cabbage volatiles. Journal of Chemical Ecology 28:131-143

Renou M, Hawlitzky N, Berthier A, Malosse C, Ramiandrasoa F (1989) Evidence for kairomones for female *Trichogramma maidis* in the eggs of the European corn borer, *Ostrinia nubilalis*. Entomophaga 34:569-580

Renou M, Nagan P, Berthier A, Durier C (1992) Identification of compounds from the eggs of *Ostrinia nubilalis* and *Mamestra brassicae* having kairomone activity on *Trichogramma brassicae*. Entomologia Experimentalis et Applicata 63:291-303

Romeis J, Babendreier D, Wäckers FL, Shanower TG (2005) Habitat and plant specificity of *Trichogramma* egg parasitoids - underlying mechanisms and implications. Basic and Applied Ecology 6:215-236

Romeis J, Shanower TG, Zebitz CPW (1997) Volatile plant infochemicals mediate plant preference of *Trichogramma chilonis*. Journal of Chemical Ecology 23:2455-2465

Salt G (1935) Experimental studies in insect parasitism. III. Host selection. Proceedings of the Royal Society of London Series B-Biological Sciences 117:413-435

Schöller M, Prozell S (2002) Response of *Trichogramma evanescens* to the main sex pheromone component of *Ephestia* spp. and *Plodia interpunctella*, (Z,E)-9,12-tetra-decadenyl acetate (ZETA). Journal of Stored Products Research 38:177-184

Schwarz HH, Huck K (1997) Phoretic mites use flowers to transfer between foraging bumblebees. Insectes Sociaux 44:303-310

Shu S, Jones RL (1989) Kinetic effect of a kairomone in moth scales of the European corn borer on *Trichogramma nubilale* Ertle & Davis (Hymenoptera: Trichogrammatidae). Journal of Insect Behavior 2:123-131

Shu S, Swedenborg PD, Jones RL (1990) A kairomone for *Trichogramma nubilale* (Hymenoptera: Trichogrammatidae). Isolation, identification, and synthesis. Journal of Chemical Ecology 16:521-529

Steidle JLM, van Loon JJA (2002) Chemoecology of parasitoid and predator oviposition behaviour. In: Hilker M, Meiners T (eds) Chemoecology of Insect Eggs and Egg Deposition. Blackwell

Publishing, Berlin, pp 291-317

Stowe MK, Turlings TCJ, Loughrin JH, Lewis WJ, Tumlinson JH (1995) The chemistry of eavesdroping, alarm and deceit. Proceedings of the National Academy of Sciences of the United States of America 92:23-28

Strand MR, Vinson SB (1982) Source and characterization of an egg recognition kairomone of *Telenomus heliothidis*; a parasitoid of *Heliothis virescens*. Physiological Entomology 7:83-90

Strand MR, Vinson SB (1983) Analyses of an egg recognition kairomone of *Telenomus heliothidis* (Hymenoptera: Scelionidae). Isolation and host function. Journal of Chemical Ecology 9: 423-432

Tabata S, Tamanuki K (1940) On the hymenopterous parasites of the pine caterpillar, *Dendrolimus sibiricus alboneatus* Mats. in southern Sakhalin. Review of Applied Entomology A 29:95

Thomson MS, Stinner RE (1990) The scale response to *Trichogramma* (Hymenoptera: Trichogrammatidae): variation among species in host specificity and the effect of conditioning. Entomophaga 35:7-21

Thornhill R (1979) Male and female sexual selection and the evolution of mating strategies in insects. In: Blum MS, Blum NA (eds) Sexual Selection and Reproductive Competition in Insects. Academic Press, New York, pp 81-121

Turlings TCJ, Gouinguené S, Degen T, Fritzsche Hoballah ME (2002) The chemical ecology of plant-caterpillar-parasitoid interactions. In: Tscharntke T, Hawkins BA (eds) Multitrophic Level Interactions. Cambridge University press, Cambridge, pp 148-173

van Huis A, Schütte C, Cools MH, Fanget P, van der Hoek H, Piquet SP (1994) The role of semiochemicals in host location by *Uscana lariophaga*, egg parasitoid of *Callosobruchus maculatus*. Proceedings of the 6th International Working Conference on Stored-Product Protection 2:1158-1164

van Lenteren JC (2000) Success in biological control of arthropods by augmentation of natural enemies. In: Gurr GM, Wratten SD (eds) Biological Control: Measures of Success. Kluwer Academic Publishers, Hingham, pp 77-103

Vet LEM, Dicke M (1992) Ecology of infochemical use by natural enemies in a tritrophic context. Annual Review of Entomology 37:141-172

Vet LEM, Lewis WJ, Carde R (1995) Parasitoid foraging and learning. In: Carde R, Bell WJ (eds) Chemical Ecology of Insects, vol 2. Chapman & Hall, New York, pp 65-101

Vet LEM, Lewis WJ, Papaj DR, van Lenteren JC (1990) A variable-response model for parasitoid foraging behavior. Journal of Insect Behavior 3:471-490

Vinson SB (1976) Host selection by insect parasitoids. Annual Review of Entomology 21:109-133

Vinson SB (1981) Habitat location. In: Nordlund DA, Jones RL, Lewis WJ (eds) Semiochemicals: Their Role in Pest Control. John Wiley & Sons, New York, pp 51-77

Vinson SB (1984) How parasitoids locate their hosts: a case of insect espionage. In: Lewis T (ed) Insect Communication. Academic Press, London, pp 325-348

Vinson SB (1985) The behavior of parasitoids. In: Kerkut GA, Gilbert LI (eds) Comprehensive Insect Physiology, Biochemistry and Pharmacology, vol 9. Pergamon Press, New York, USA, pp 417-469

Vinson SB (1991) Chemical signals used by parasitoids. In: Bin F (ed), vol. 74, 3 edn, Perugia, pp 15-42

Wäckers FL (2005) Suitability of (extra-)floral nectar, pollen, and honeydew as insect food resources. In: Wäckers FL, van Rijn PCL, Bruin J (eds) Plant-Provided Food for Carnivorous Insects: A Protective Mutualism and its Applications, Cambridge University Press, Cambridge, pp 17-74

Wegener R, Schulz S (2002) Identification and synthesis of homoterpenoids emitted from elm leaves after elicitation by beetle eggs. Tetrahedron 58:315-319

Wegener R, Schulz S, Meiners T, Hadwich K, Hilker M (2001) Analysis of volatiles induced by oviposition of elm leaf beetle *Xanthogaleruca luteola* on *Ulmus minor*. Journal of Chemical Ecology 27:499-515

Weseloh RM (1981) Host location in parasitoids. In: Nordlund DA, Jones RL, Lewis WJ (eds) Semiochemicals: Their Role in Pest Control. John Wiley & sons, New York, pp 79-95

Winkler K, Wäckers FL, Stingli A, van Lenteren JC (2005) *Plutella xylostella* (diamondback moth) and its parasitoid *Diadegma semiclausum* show different gustatory and longevity responses to a range of nectar and honeydew sugars. Entomologia Experimentalis et Applicata 115: 187-192

Yasuda K, Tsurumachi M (1995) Influence of male-adults of the leaf-footed plant bug, *Leptoglossus australis* (Fabricius) (Heteroptera, Coreidae), on host-searching of the egg parasitoid, *Gryon pennsylvanicum* (Ashmead) (Hymenoptera, Scelionidae). Applied Entomology and Zoology 30:139-144

Yoshimoto CM (1976) *Pseudoxenufens forsynthi* a new genus and species of Trichogrammatidae (Hymenoptera: Chalcidoidea) from western Ecuador. Canadian Entomologist 108:419-422

Zaborski E, Teal PEA, Laing JE (1987) Kairomone-mediated host finding by spruce budworm egg parasite, *Trichogramma minutum*. Journal of Chemical Ecology 13:113-122

Zaki FN (1985) Reactions of the egg parasitoid *Trichogramma evanescens* Westw. to certain insect sex pheromones. Zeitschrift für Angewandte Entomologie 99:448-453

Zuk M, Kolluru GR (1998) Exploitation of sexual signals by predators and parasitoids. The Quarterly Review of Biology 73:415438

3 | Chemical Communication: Butterfly Anti-Aphrodisiac Lures Parasitic Wasps

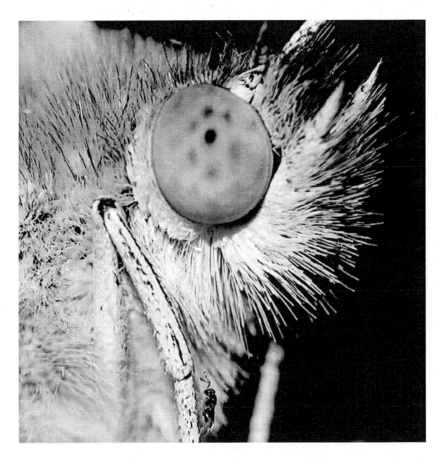

N.E. Fatouros, M.E. Huigens, J.J.A. van Loon, M. Dicke & M. Hilker

Chapter 3

Chemical Communication:
Butterfly Anit-Aphrodisiac Lures Parasitic Wasps

Abstract

Parasitic insects display a tremendous variety of host location strategies. Here, we describe a novel strategy of parasitic wasps exploiting the intraspecific communication system of herbivorous insect hosts, i.e. chemical espionage on an anti-aphrodisiac. This chemical cue, passed from male to female butterflies during mating, renders females less attractive to conspecific males. Upon detection of the anti-aphrodisiac odour, the tiny parasitic wasp *Trichogramma brassicae* employs phoresy and is transported to the eggs of its host, the large cabbage white butterfly. A wasp first mounts a mated female butterfly marked with the anti-aphrodisiac, and thereafter hitch-hikes with her to a host plant. When the butterfly starts to lay eggs, the wasp descends onto the leaf and parasitises the females' freshly laid eggs. To our knowledge, these results are the first to show chemical espionage of anti-aphrodisiacs by parasitic wasps, which can lead to a dramatic reduction of host offspring survival. This fascinating strategy may have evolved frequently in nature and adds a new dimension to our understanding of host-parasitoid associations.

Keywords: host location, anti-sex pheromone, phoresy, *Trichogramma brassicae*, *Pieris brassicae*

Introduction

In order to locate tiny eggs of herbivorous host insects in an ocean of plant biomass, parasitic wasps have been shown to employ chemical cues a) from the plant induced by host egg-deposition (Fatouros et al. 2005; Hilker and Meiners 2002) or b) from the adult host stage (i.e. infochemical detour) (Vet and Dicke 1992) rather than cues emanating from the eggs themselves.

Eavesdropping on sex pheromone signals from the host have been shown for several egg parasitoid species (Powell 1999). In contrast, anti-aphrodisiacs have so far been neglected in research on chemical espionage. These pheromones evolved in males to enhance their post-mating success by decreasing the attractiveness of females to conspecific mates (Gilbert 1976; Happ 1969).

We hypothesised that anti-aphrodisiacs are more reliable cues indicating future host egg production than sex pheromones. We therefore investigated whether anti-aphrodisiacs can be perceived by egg parasitoids. Moreover, we revealed phoretic behaviour by parasitic wasps ensuing perception of a host's anti-aphrodisiac. Phoresy constitutes a directed and highly efficient host locating mechanism. It is defined as the transport of certain insects on the bodies of others for purposes other than direct parasitism (Howard 1927). Tiny parasitic wasps have limited control over their flight direction and would derive an adaptive benefit by exploiting an anti-aphrodisiac, followed by phoresy on a mated adult female host. The association between butterflies and their egg parasitoids provides an excellent opportunity to verify the existence of such a strategy.

The minute, ca. 0.5 mm small, parasitic wasp *Trichogramma brassicae* Bezdenko (Hymenoptera: Trichogrammatidae) occurs in temperate zones where it parasitises eggs of the large cabbage white *Pieris brassicae* L. (Lepidoptera: Pieridae) and other related species (Pak et al. 1990; van Heiningen et al. 1985). The gregarious butterfly species lays clutches of more than 20 eggs on several *Brassica* species (Feltwell 1982). A female *T. brassicae* wasp can parasitise 15 or more eggs of *P. brassicae* and typically, 2 - 4 wasps emerge from a single host egg (N.E. Fatouros, personal observations).

Recently, *P. brassicae* males were shown to synthesise an anti-aphrodisiac, benzyl cyanide, that is transferred to the females within their ejaculate (Andersson et al. 2003). After mating, female butterflies smell like males for several days while females do not produce this compound themselves. Here, we tested whether *T. brassicae* uses this anti-aphrodisiac to locate mated *P. brassicae* females, thereafter hitch-hiking along to parasitise her newly laid eggs.

Methods

Insects. *Trichogramma brassicae* (strain Y175) wasps originated from a cabbage field in the Netherlands and were reared on eggs of *Pieris brassicae* for several generations after collection. Only mated, 2 - 5 day-old naïve female wasps (no contact to eggs prior to the experiment) were used for the experiments, with the exception of the flight chamber test where oviposition-experienced females were used.

   Pieris brassicae adults were obtained from a laboratory colony maintained on Brussels sprout plants (*Brassica oleraceae* L. var. *gemmifera* cv. Cyrus). Mated females and males were obtained by taking copulating pairs from the rearing. Virgin females and males were separated in the pupal stage (Feltwell 1982).

Response to butterfly odours. The experiments were carried out in a two-chamber olfactometer (Fatouros et al. 2005). Two adult butterflies per chamber were introduced as the odour source. The time spent by the wasps in one of the two odour fields was observed for 300 seconds. Each day 10-15 naïve wasps were tested, for a total of 40 wasps per combination. To avoid biased results due to possible side preferences of the parasitoids, the olfactometer was rotated 180° after every third insect.

Response to butterfly anti-aphrodisiac. In the same two-chamber olfactometer, the response of naïve *T. brassicae* towards the anti-aphrodisiac (benzyl cyanide) of mated *P. brassicae* females was tested. A virgin female was painted with 10 µl of a benzyl cyanide (Aldrich, purity 99%) solution in hexane and tested against a virgin female treated with 10 µl of hexane. Five different doses of benzyl cyanide were tested: a) 1.015 µg/µl (10^{-3} dilution), b) 0.203 µg/µl (5×10^{-3} dilution), c) 0.1015 µg/µl (10^{-4} dilution), d) 0.02 µg/µl (5×10^{-4} dilution), and e) 0.01 µg/µl (10^{-5} dilution). After each 5th wasp, the test and control butterflies were replaced. A total of 40 wasps was tested per dose.

Mounting on butterfly. Mounting behaviour of *T. brassicae* females was tested in a two-choice bioassay conducted in a plastic container (9 cm high, 13.5 cm diameter). Two adults of *P. brassicae* were placed in the arena after cooling down in a refrigerator to decrease mobility. A naïve *T. brassicae* female was introduced and continuously observed until it climbed onto one of the two butterflies. The body part on which the female wasp mounted was recorded. When a wasp did not mount one of the two butterflies within 5 minutes, a "no response" was recorded. After each 10th wasp, the butterflies were replaced. For each combination, 35 wasps were tested.

An additional mounting experiment tested two virgin females, one painted with 10 µl of the anti-aphrodisiac benzyl cyanide dissolved in hexane (0.203 µg/µl) and the other painted with 10 µl hexane. The experimental conditions were the same as described above, except for the following: both butterflies were placed under a small gauze mesh (3 x 6 cm, 2 mm mesh size) to prevent them from moving and to exclude visual cues. A naïve *T. brassicae* female was introduced and continuously observed until it went into a cage and mounted the butterfly. Butterflies were treated with 10 µl of benzyl cyanide solution and hexane, respectively, every 10 minutes. Ten wasps were tested on each pair of virgin butterflies and a total of 35 wasps were tested.

Phoresy on butterflies. Experiments were conducted in a flight chamber installed in a greenhouse compartment under the following conditions: 25 ± 3°C, 50-80 % RH. In a glasshouse compartment (4 x 2.5 x 3.5 m), a tent (3.3 x 1.8 x 3 m) made of white sheets was constructed. No directed airflow was used, but a turbulent airflow inside the tent was present as the greenhouse compartment was continuously ventilated. Observations were made under natural daylight.

A Brussels sprouts plant, 8-12 weeks old, was placed on one end of the table. A mated *P. brassicae* female with a high egg load was offered to an oviposition-experienced female egg parasitoid in a small arena on the other end of the table approximately 1.5 meter away. Both the butterfly and the parasitoid were observed and their behaviour recorded using The Observer software 3.0 (Noldus Information Technology 1993[©]). When a *Pieris* female carrying a *T. brassicae* female did not land on the cabbage plant after take-off, it was scored as a non-responder. When a *Pieris* female landed on the plant and started to oviposit, the position and movement of the parasitoid was scored using the following descriptors: a) lost (parasitoid not found back on the butterfly), b) reaches the host habitat (parasitoid was found on butterfly but lost trying to descend onto the plant), c) reaches host plant (parasitoid climbed onto the plant but was lost thereafter), and d) observed parasitism (parasitoid found butterfly eggs and parasitized them during the observation).

Results

The first objective was to determine whether *T. brassicae* wasps are able to detect adult *P. brassicae* butterflies from a distance. In two-choice olfactory bioassays the wasps showed clear preferences for odours of mated *P. brassicae* females or males (Figure 1). Wasps were significantly more strongly arrested by the scent of mated than by the scent of virgin females. When odour from virgin female butterflies was offered against odour from male butterflies the wasps preferred volatiles from the male butterflies. The wasps did not, however, discriminate between male and mated female butterflies. When offered separately against clean air, odours from mated female or male butterflies significantly arrested the wasps, but odours from virgin females did not.

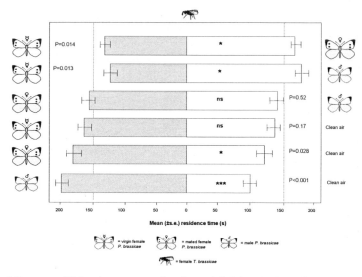

Figure 1: Response of *T. brassicae* wasps to odour from adult *P. brassicae* butterflies. Mean residence time (± s.e.) in test and control field of a two-chamber olfactometer are given. N=40 tested wasps per experiment. Asterisks indicate significant differences *=P<0.05, ***=P<0.001, n.s. not significant (Wilcoxon's matched pairs signed ranks test).

The second objective was to determine at which concentration the anti-aphrodisiac benzyl cyanide (Andersson et al. 2000; 2003) attracts the wasp *T. brassicae*, using similar behavioural trials as described above. The wasps were significantly arrested by odour from virgin female butterflies treated with 10 µl of a benzyl cyanide solution containing 0.203 µg/µl (Figure 2; P=0.008, Wilcoxon's matched pairs signed rank test) and 0.102 µg/µl (Figure 2; P=0.016, Wilcoxon's matched pairs signed rank test) when solvent-treated virgin females were offered as alternative. This result shows that *T. brassicae* uses the anti-aphrodisiac benzyl cyanide from *P. brassicae* to discriminate between mated and unmated females.

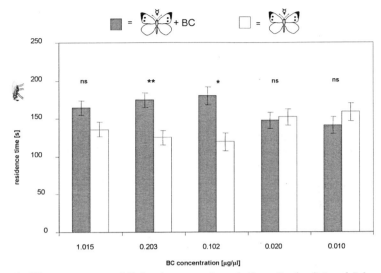

Figure 2: Olfactory response of *T. brassicae* wasps towards the anti-aphrodisiac of *P. brassicae* butterflies. Virgin females butterflies were painted with benzyl cyanide (BC) diluted in hexane (grey bars) or with hexane alone (white bars). Mean residence time (± s.e.) in test and control fields of a two-chamber olfactometer are given. N=40 tested wasps. Asterisks indicate significant differences *=$P<0.05$, **=$P<0.01$, n.s. not significant (Wilcoxon's matched pairs signed ranks test).

Subsequently, we investigated whether the butterfly anti-aphrodisiac results in the wasp mounting a mated adult *P. brassicae* female and in using her as a transport vehicle. We exposed adult butterflies to female wasps in two-choice bioassays. Indeed, the wasps mounted adult butterflies. Moreover, the wasps significantly preferred climbing onto mated females over virgin female butterflies (Figure 3). When offering a mated female against a male butterfly, the wasps mounted the mated female butterfly significantly more frequently. Virgin butterfly females were by far the least attractive specimens for the wasps to climb on. Mounting occurred most frequently on the butterfly's wings (73 %, $P<0.001$, binomial distribution probability test).

In additional tests, virgin female butterflies painted with benzyl cyanide diluted in hexane (0.203 µg/µl) were offered against virgin female butterflies painted with only hexane. The wasps preferred to mount virgin female butterflies treated with the anti-aphrodisiac significantly more often than those treated with the solvent control (Figure 3). Clearly, the presence of the anti-aphrodisiac induces the wasps

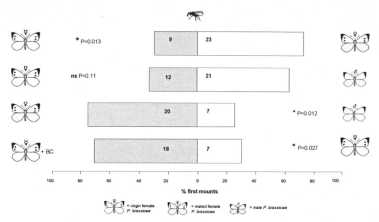

Figure 3: Percent of first mounts by *T. brassicae* wasps on adult *P. brassicae* butterflies. Number of responding *T. brassicae* shown inside the bars. N=40 wasps tested per combination. BC = benzyl cyanide. Asterisks indicate significant differences within the choice test, *=P<0.05, n.s.= not significant (Chi²-Test).

to climb onto an adult mated female butterfly.

Finally, to determine whether *T. brassicae* wasps are able to remain on an adult female butterfly during flight and subsequently parasitise her freshly laid eggs, we conducted bioassays in a flight chamber. In total, 28 mated female butterflies each carrying a single wasp were observed to fly towards a Brussels sprout plant. Fourteen wasps (50%) reached the host habitat, i.e. were found on the butterfly after it landed on the plant and started laying eggs. Four of these wasps reached the host plant by descending from the butterfly onto the cabbage plant, and two of them were observed to parasitise the freshly laid eggs after leaving the butterfly. Thus, phoresy on an adult *P. brassicae* female led to successful parasitism by *T. brassicae* in at least 7.1% of the observations.

Discussion

Our results report in detail the sophisticated combination of chemical espionage with a phoretic transportation strategy, leading from detection of a host's chemical cue to parasitism. We demonstrate that *T. brassicae* females spy on the anti-aphrodisiac of *P. brassicae* butterflies, benzyl cyanide, to find and subsequently use their most profitable adult transport vehicle: a mated female marked with this pheromone. Such a strategy may be a widespread phenomenon in egg parasitoids. Because anti-aphrodisiacs are species-specific cues, *T. brassicae* is likely to be an opportunistic specialist on *P. brassicae* in nature, something that is also expected for certain other *Trichogramma* spp. (Pak et al. 1990) *Pieris brassicae* butterflies are highly mobile insects. Flowers may serve as likely mounting locations for *Trichogramma*, as shown for mites phoretic on bumble bees (Schwarz and Huck 1997). For *Trichogramma* spp., as for most parasitoid species, sugars obtained from floral nectar are an essential energy source. Flower location in these parasitoids is mediated by both olfactory and visual stimuli (Romeis et al. 2005). Thus, since the butterfly host and *Trichogramma* wasps both use floral nectar, flowers may be used for horizontal transfer. On the practical side, the butterfly is stationary during nectar feeding, making it easier to be mounted by the wasp.

We found that neither volatiles from virgin females, nor contact with them elicited a positive response in *T. brassicae*. An earlier study had demonstrated that odours emitted by virgin *P. brassicae* females when in a 'calling position' attracted *Trichogramma evanescens* (Noldus 1989). We rarely observed this posture of virgin *P. brassicae* females in our experiments. It is unknown which substance(s) are emitted by calling virgin females. Andersson et al. (2003) have shown that virgin females release volatiles qualitatively different from those released by mated females. These compounds might constitute a sex pheromone. However, mate location in butterflies is assumed to be guided principally by visual stimuli (Obara 1970; Sweeney et al. 2003). The discrepancy between the response of *T. evanescens* and *T. brassicae* towards *P. brassicae* suggests that different parasitoid species can exploit different cues released by the same host species.

Information on phoresy by *Trichogramma* spp. in nature is scarce. The only published report is of *Trichogramma* females found on a moth species in Russia (Tabata and Tamanuki 1940). Besides this, we have recently found *Trichogramma* females on adult *P. brassicae* females in The Netherlands (N.E. Fatouros, personal observations). It is obvious that field studies carefully monitoring adult butterflies and moths for the presence of *Trichogramma* wasps are needed. Field data on parasitism rates of more than 50 % revealed that *Trichogramma* wasps are successful in finding *Pieris brassicae* eggs in nature (van Heiningen et al. 1985).

Phoresy is common among some Trichogrammatidae and Scelionidae (Clausen 1976; Vinson 1998). However, very little is known regarding the host location behaviour of phoretic egg parasitoids. The little available information concerns scelionids, which use sex or aggregation pheromones to locate adult hosts (Aldrich et al. 1984; Arakaki et al. 1996). The egg parasitoid *Telenomus euproctidis* was more attracted to virgin females of the tussock moth than to mated ones. In this case, two compounds of the moth's sex pheromone were shown to attract the egg parasitoid

(Arakaki et al. 1996). Obviously, female moth sex pheromones are exploited by phoretic egg parasitoids but we expect the utilization of anti-aphrodisiacs to occur more frequently. Hitch-hiking with a virgin female or a male host is expected not to be as efficient in locating host eggs as performing phoresy on a mated female host. Moreover, in the case of moths, it may be difficult for *Trichogramma* spp., which are diurnal egg parasitoids, to encounter adult female moths since they generally release their sex pheromones during the scotophase. Sex pheromones of moth species like *Heliothis zea* and *Mamestra brassicae* have been shown to elicit a response in *Trichogramma* species (Noldus 1989; Noldus et al. 1991a; Noldus et al. 1991b). In some cases, trace amounts of the moths' sex pheromones were found to be absorbed onto plant surfaces or to scales and subsequently re-released thus acting as a bridge in time and space for the chemical spies (Arakaki and Wakamura 2000; Noldus et al. 1991a).

The fascinating host location strategy of chemical espionage on an anti-aphrodisiac in combination with phoresy on a mated female host may have evolved frequently in egg parasitoids. From an egg parasitoids point of view, this strategy is most adaptive in species which a) have limited flight ability, b) have only a narrow window of time to successfully parasitize eggs, and/or c) parasitise gregarious hosts. In addition to limited capability for directed flight under natural conditions, *Trichogramma* wasps generally parasitize younger host eggs more successfully than older ones. Sclerotization of the head capsule of the lepidopteran embryo may mark the end of successful parasitism (Pak 1986). Some egg parasitoids seem to have a very narrow time window available as they can only gain offspring from freshly laid host eggs (Orr et al. 1986).

In general, it is most adaptive for an egg parasitoid to be phoretic on a host that lays a clutch of eggs. In this case, the parasitoid has ample time to descend from the adult female host onto the leaf, locate the host eggs and, because of the high egg number, produce many offspring. From the perspective of the host, egg parasitoids are significant mortality factors and consequently will exert selection on butterfly intraspecific communication. Exploitation of the anti-aphrodisiac by egg parasitoids is therefore expected to seriously constrain its evolution. Future studies into genetic variation in this vital trait linked to local parasitoid selection pressure hold good promise.

Acknowledgements

The authors are grateful to Barbara Randlkofer for assistance with the experiments; Jeffrey A. Harvey and Suzanne Blatt for discussion and comments on an early draft of the manuscript.

Gabriella Bukovinszkine'Kiss, Leo Koopman, Frans van Aggelen, and André Gidding are thanked for culturing the insects. This work was supported by the Deutsche Forschungsgemeinschaft (Hi 416 / 15-1,2).

References

Aldrich JR, Kochansky JP, Abrams CB (1984) Attractant for beneficial insect and its parasitoids: pheromone of the predatory spined soldier bug, *Podisus maculiventris* (Hemiptera: Pentatomidae). Environmental Entomology 13:1031-1036

Andersson J, Borg-Karlson A-K, Wiklund C (2000) Sexual cooperation and conflict in butterflies: a male-transferred anti-aphrodisiac reduces harassment of recently mated females. Proceedings of the Royal Society: Biological Sciences 267:1271-1275

Andersson J, Borg-Karlson A-K, Wiklund C (2003) Antiaphrodisiacs in Pierid butterflies: a theme with variation! Journal of Chemical Ecology 29:1489-1499

Arakaki N, Wakamura S (2000) Bridge in time and space for an egg parasitoid-kairomonal use of trace amount of sex pheromone adsorbed on egg mass scale hair of the tussock moth, *Euproctis taiwana* (Shiraki) (Lepidotera: Lymantriidae), by an egg parasitoid, *Telenomus euproctidis* Wilcox (Hymenoptera: Scelionidae), for host location. Entomological Science 3: 25-31

Arakaki N, Wakamura S, Yasuda T (1996) Phoretic egg parasitoid, *Telenomus euproctidis* (Hymenoptera: Scelionidae), uses sex pheromone of tussock moth *Euproctis taiwana* (Lepidoptera: Lymantriidae) as a kairomone. Journal of Chemical Ecology 22:1079-1085

Clausen CP (1976) Phoresy among entomophagous insects. Annual Review of Entomology 21:343-368

Fatouros NE, Bukovinszkine'Kiss G, Kalkers LA, Soler Gamborena R, Dicke M, Hilker M (2005) Oviposition-induced plant cues: do they arrest *Trichogramma* wasps during host location? Entomologia Experimentalis et Applicata 115:207-215

Feltwell J (1982) Large White Butterfly - The Biology, Biochemistry and Physiology of *Pieris brassicae* (Linnaeus). Dr. W. Junk Publishers, The Hague-Boston-London

Gilbert LE (1976) Postmating female odor in *Heliconius* butterflies: A male-contributed antiaphrodisiac? Science 193:419-420

Happ GM (1969) Multiple sex pheromones of the mealworm beetle, *Tenebrio molitor* L. Nature 222: 180-181

Hilker M, Meiners T (2002) Induction of plant responses towards oviposition and feeding of herbivorous arthropods: a comparison. Entomologia Experimentalis et Applicata 104:181-192

Howard LO (1927) Concerning phoresy in insects. Entomological News 38:145-147

Noldus LPJJ (1989) Semiochemicals, foraging behaviour and quality of entomophagous insects for biological control. Journal of Applied Entomology 108:425-451

Noldus LPJJ, Potting RPJ, Barendregt HE (1991a) Moth sex pheromone adsorption to leaf surface: bridge in time for chemical spies. Physiological Entomology 16:329-344

Noldus LPJJ, van Lenteren JC, Lewis WJ (1991b) How *Trichogramma* parasitoids use moth sex pheromones as kairomones: orientation behaviour in a wind tunnel. Physiological Entomology 16:313-327

Obara Y (1970) Studies on mating behavior of the white cabbage butterfly, *Pieris rapae crucivora* Boisduval III. Near-ultra-violet reflection as signal of intraspecific communication. Zeitschrift für vergleichende Physiologie 69:99-116

Orr DB, Russin JS, Boethel DJ (1986) Reproductive biology and behavior of *Telenomus calvus* (Hymenoptera: Scelionidae), a phoretic egg parasitoid of *Podisus maculiventris* (Hemiptera: Pentatomidae). Canadian Entomologist 118:1063-1072

Pak GA (1986) Behavioural variations among strains of *Trichogramma spp.* A review of the literature on host-age selection. Journal of Applied Entomology 101:55-64

Pak GA, Kaskens JWM, Jong de EJ (1990) Behavioural variation among strains of *Trichogramma spp.*: host-species selection. Entomologia Experimentalis et Applicata 56:91-102

Powell W (1999) Parasitoid hosts. In: Hardie J, Minks AK (eds) Pheromones of Non-Lepidopteran Insects Associated with Agricultural Plants. CABI Publishing, Wallingford, pp 405-427

Romeis J, Babendreier D, Wäckers FL, Shanower TG (2005) Habitat and plant specificity of *Trichogramma* egg parasitoids - underlying mechanisms and implications. Basic and Applied Ecology 6:215-236

Schwarz HH, Huck K (1997) Phoretic mites use flowers to transfer between foraging bumblebees. Insectes Sociaux 44:303-310

Sweeney A, Jiggins C, Johnson S (2003) Polarized light as a butterfly mating signal. Nature 423:31

Tabata S, Tamanuki K (1940) On the hymenopterous parasites of the pine caterpillar, *Dendrolimus sibiricus alboneatus* Mats. in southern Sakhalin. Review of Applied Entomology A 29:95

van Heiningen TG, Pak GA, Hassan SA, van Lenteren JC (1985) Four year's results of experimental releases of *Trichogramma* egg parasites against lepidopteran pests in cabbage. Mededelingen van de Faculteit Landbouwwetenschappen Rijksuniversiteit Gent 50:379-388

Vet LEM, Dicke M (1992) Ecology of infochemical use by natural enemies in a tritrophic context. Annual Review of Entomology 37:141-172

Vinson SB (1998) The general host selection behavior of parasitoid Hymenoptera and a comparison of initial strategies utilized by larvaphagous and oophagous species. Biological Control 11: 79-96

4 Oviposition-Induced Plant Cues: Do They Arrest *Trichogramma* Wasps During Host Location?

N.E. Fatouros, G. Bukovinszkine'Kiss, L.A. Kalkers, R. Soler Gamborena,
M. Dicke & M. Hilker

Chapter 4

Oviposition-Induced Plant Cues: Do They Arrest *Trichogramma* Wasps During Host Location?

Abstract

Plants may defend themselves against herbivorous insects already before larvae hatch from the eggs and start feeding. One of these preventive defence strategies is to produce plant volatiles in response to egg deposition, which attract egg parasitoids killing the herbivore eggs. Here, we studied whether egg deposition by *Pieris brassicae* L. (Lepidoptera: Pieridae) induces Brussels sprouts plants to produce cues that attract or arrest *Trichogramma brassicae* Bezdeko (Hymenoptera: Trichogrammatidae). Olfactometer bioassays revealed that odours from plants with eggs did not attract or arrest parasitoids. However, contact bioassays showed that *T. brassicae* females were arrested on egg-free leaf squares excised from leaves with 72 h-old egg masses, which are highly suitable for parasitisation. We tested the hypothesis that this arresting activity is due to scales and chemicals deposited by the butterflies during oviposition and thus present on the leaf surface in the vicinity of eggs. Indeed, leaf squares excised from egg-free leaves, but contaminated with butterfly deposits, arrested the wasps when the squares were tested 24 h after contamination. However, squares from egg-free leaves with 72 h-old butterfly deposits had no arresting activity. Thus, we exclude that the arresting activity of leaf area near to 72 h-old egg masses was elicited by cues from scales and other butterfly deposits. We suggest that egg deposition of *P. brassicae* induces a change of leaf surface chemicals in leaves with egg masses. A systemic induction extending to an egg-free leaf neighboured to an egg-carrying leaf could not be detected. Our data suggest that a local, oviposition-induced change of leaf surface chemicals arrests *T. brassicae* in the vicinity of host eggs.

Keywords: egg parasitoids, oviposition, tritrophic interactions, host finding, herbivore-induced volatiles, synomones, Brussels sprouts, *Pieris brassicae*, *Trichogramma brassicae*

Introduction

Trichogramma wasps are minute egg parasitoids of especially lepidopteran species. Because of their widespread application as biological control agents, their host selection behaviour has been intensively investigated (reviewed by Noldus 1989b; Wajnberg and Hassan 1994). Host selection is divided into three steps: host-habitat location, host location and host acceptance (Nordlund et al. 1988). The use of infochemicals plays a major role during host selection by parasitic wasps in general, including *Trichogramma* (Lewis and Martin 1990; Nordlund 1994; Schmidt 1994; Vinson 1976).

In the process of host location, usually chemical cues from the host itself (i.e. kairomones) are exploited by the parasitoids (overview in Wajnberg and Hassan 1994). Egg parasitoids like *Trichogramma* spp. have to cope with the problem of searching for an immobile and inactive host stage. Therefore, they often rely on infochemicals from other stages such as the host adult that are easier to detect and additionally give a high probability for the presence of the host eggs (i.e. infochemical detour) (Vet and Dicke 1992). Non-volatile kairomones such as wing scales from lepidopteran adult hosts affect the searching behaviour of *Trichogramma* spp. resulting in increased chances of encountering host eggs (see review by Noldus 1989b). In addition, volatile infochemicals, e.g. sex pheromones from calling moths, can have a kairomonal effect (see Boo and Yang 2000; Reddy et al. 2002; Schöller and Prozell 2002 and references therein). The eggs themselves can be a direct source of volatile or contact kairomones as well (see Boo and Yang 2000 and references therein).

Moreover, plants produce chemical stimuli that can mediate host habitat location behaviour in parasitic wasps. In the past, especially feeding-induced plant volatiles exploited by larval parasitoids have been studied (reviewed by Dicke 1999). However, plants may not only respond towards feeding by herbivores, but already earlier to the oviposition by herbivorous insects, thus activating a preventive anti-herbivore defence strategy. For three tritrophic systems oviposition by herbivorous insects was shown to induce a change in the emission of plant volatiles resulting in the attraction of egg parasitoids such as *Oomyzus gallerucae*, *Chrysonotomyia ruforum*, or *Trissolcus basalis* (Colazza et al. 2004a; reviewed by Hilker and Meiners 2002). It has not yet been tested whether generalist egg parasitoids such as *Trichogramma* species respond to oviposition-induced changes of the plant's chemistry.

In the present study we investigated a tritrophic system consisting of Brussels sprouts (*Brassica oleracea* var. *gemmifera* cv. Cyrus), the large cabbage white butterfly

Pieris brassicae L. (Lepidoptera: Pieridae), and the egg parasitoid *Trichogramma brassicae* Bezdeko (Hymenoptera: Trichogrammatidae). Egg deposition by *P. brassicae* has been suggested to induce changes in plant surface chemistry of Brussels sprout plants (Blaakmeer et al. 1994a).

We hypothesized that egg deposition by *P. brassicae* induces changes in the emission of cabbage leaf volatiles or in chemicals on the plant leaf surface that are exploited by *T. brassicae* to locate host eggs. First, the arrestment time of *T. brassicae* on cabbage leaves with egg masses was compared to the time spent on leaves without eggs. Further bioassays addressed the question whether the longer arrestment of oviposition-experienced *T. brassicae* on egg-laden leaves was due to cues from the eggs or from butterfly deposits such as scales or butterfly odour adsorbed to the surface. When it turned out that neither cues from eggs nor from butterfly deposits were necessary to arrest the wasps, we studied whether egg deposition induced a change of plant cues (a) in the vicinity of egg masses (local induction), and (b) in egg-free leaves adjacent to egg-laden ones (systemic induction) increased the wasps' response. Furthermore, their response to volatiles from egg-laden leaves was tested in both dynamic and static olfactometer bioassays.

Material and methods

Plants and herbivores. Brussels sprout plants were reared in a greenhouse (18 ± 2°C, 70% rh, L16:D8). Plants of 8-12 weeks having ca. 14-16 leaves were used for rearing of *Pieris brassicae* and for the experiments.

Pieris brassicae was reared on Brussels sprout plants in a climate room (21 ± 1°C, 50-70% r.h., L16:D8). Each day a plant was placed into a large cage (80 x 100 x 80 cm) with more than 100 adults for approximately 24 hours to allow egg deposition.

Parasitoids. *Trichogramma brassicae* (strain Y175) was reared in eggs of *P. brassicae* for more than 60 generations (25°C, 50-70% rh, L16:D8). For the rearing, 1-3 day-old *P. brassicae* eggs on leaves were used. Initially mated, naïve wasps (no oviposition experience in eggs of *P. brassicae*) were used (contact bioassay no. 1a, compare Table 1). All following bioassays were done with mated, oviposition-experienced female wasps, because naïve wasps were shown to have a low response level. However, for the host-age suitability tests naïve wasps were used again. An oviposition experience was given for a period of 18 h prior to the experiment with 2-3 day-old *P. brassicae* eggs deposited on Brussels sprout leaves. Wasps were about 2-5 days old when tested. They were always provided with honey solution prior to the experiment.

Contact bioassays with leaf squares. For the experiments, test plants were placed into the cage with more than 100 *P. brassicae* adults to allow deposition of eggs, wing scales, and host odours onto the plants. After this exposure during generally 8 h to the butterflies, the treated plants were tested immediately or were kept in a climate chamber (21 ± 2°C, 70% rh, L16:D8) either just overnight until the next day or for another 24 to 72 h after the exposure day. Thus, the period during which eggs or butterfly deposits could affect and induce the cabbage plant was in total either less than 12 h, about 24 h, 48 h, 72 h or 96 h (Compare Table 1). Control plants were never in contact with *P. brassicae* or any other insect. However, they were grown under the same abiotic conditions as treated plants.

For the experiments, only turgid leaves were used. Always two leaves of corresponding size and position relative to the topmost leaves were used. Corresponding sections were cut from the control and the treated leaf, so that size and structure were as similar as possible.

A wasp was released in the centre of a small glass Petri dish (5.5 cm diameter) lined with filter paper. The wasps were simultaneously offered a test and a control leaf square of 2 cm^2 cut from an excised leaf. The total duration of time spent on the leaves (with antennal drumming during most of the time) was observed for a period of 300 sec using The Observer software 3.0 (Noldus Information Technology 1993©). The time spent searching in the area outside the leaf squares was scored as "no response". The number of wasps per experiment ranged from 30 to 64 (indicated in figure legends). Test and control squares were changed after having tested 3 wasps. Each wasp was used only once and was then discarded.

Bioassays (Compare Table 1). In contact experiment (1) denoted by "eggs", a leaf square with an egg clutch (≤ 50 eggs) (=test) was tested against a "clean" leaf square taken from an egg-free control plant (a) with naïve T. brassicae females and (b) with oviposition-experienced ones. The test leaf squares carried eggs, which had been laid about 48 h prior to the experiment. All following experiments were conducted with oviposition experienced T. brassicae females. In contact experiment (2) denoted by "eggs removed", the egg mass laid on a test square was carefully removed with a brush just prior to the bioassay. Such a test square was tested against a "clean" leaf square of an egg-free control plant. In contact experiment (3) denoted by "deposits", leaf squares were cut from a plant that had never received eggs, but had been explored by butterflies indicated by wing scales deposited on the leaves.

Table 1. An overview of the experiments

Experiment No.		Duration of leaf treatment [h]	Treatment
		Contact bioassays	
Contact 1	a	48	Eggs (naïve wasps)
	b	48	Eggs
Contact 2		72	Eggs removed
Contact 3	a	24	Deposits
	b	72	Deposits
Contact 4	a	<12, 24, 48, 72, 96	Locally induced (short range)
	b	72	Locally induced (medium range)
Contact 5		72	Systemic induced
Contact 6		96	Synergistic effect
		Olfactometer bioassays	
Y-tube		24-72	Egg-leaf vs. clean air (long range)
Two-chamber 1		72	Egg-leaf vs. clean air (short range)
Two-chamber 2		72	Egg-leaf vs. clean leaf (short range)

Egg deposition onto the plants was prevented by covering the plants with gauze (2 mm^2 mesh size). Test leaf squares (a) ca. 24 h, or (b) ca. 72 h after exposure to the butterflies were tested against "clean" control squares. Control plants were equally covered for the same period, but were kept separate from butterflies so that no butterfly odour or scales could adsorb. In contact experiment (4) denoted by "locally induced", the response of T. brassicae to a section of leaf area in the vicinity of an egg mass was tested. Test leaf squares were taken (a) in the very close vicinity of an egg mass (short range), and (b) about 5 cm away from an egg mass (medium range). When test leaf squares were taken from short range, egg masses had been deposited about 12 h, 24 h, 48 h, 72 h, or 96 h prior to the bioassay. These test squares were tested against "clean" leaf squares. When test leaf squares were taken from medium range, the eggs had been deposited about 72 h prior to the bioassay. In contact experiment (5) denoted by "systemically induced", a plant of which one leaf was covered by gauze (about 0.5 mm^2 mesh size) was placed into a butterfly cage for 8 h. The gauze prevented egg deposition on the covered leaf, whereas eggs could be laid onto all other leaves of the plant. After being exposed, about 1-5 egg masses per leaf had been laid onto the test plants. Leaves from control plants were equally covered for the same period, but these plants were kept apart from butterflies. About 72 h after egg deposition on the test plants, squares were cut from the covered test leaves and from leaves of a control plant and were used in the bioassay. In contact experiment (6) denoted by "synergistic effect", test leaf squares were taken from a leaf carrying eggs and damaged by larval feeding. On an egg-laden plant with 72 h-old eggs, ca. 50 first instar larvae were placed and allowed to feed on it for another 48 h. Squares cut from a leaf area without feeding damage and about 3-5 cm away from the egg mass were tested against leaf squares from "clean" control plants.

Dynamic Y-tube olfactometer test. To test whether the wasps use volatiles to locate leaves carrying egg batches, 2-choice bioassays were conducted in a Y-tube olfactometer. The olfactometer has been described by Takabayashi and Dicke (1992). It was made of glass (4 cm ID; stem 6 cm, arms 10 cm; stem-arms angle 60°) with each arm connected to a glass container (2 l) holding the odour source. Air was filtered through activated charcoal, humidified and split into two air streams that were led through the glass containers to the olfactometer with a flow of 1 l/min in each arm. At the end of the stem it was sucked through with a flow of 2 l/min. All experiments were conducted at 20-25°C and 50-60% relative humidity, using a light bulb above the olfactometer (Philips, The Netherlands, HPL Comfort, 50 W). A cardboard box (15 cm high, 40 cm wide) surrounded the Y-tube olfactometer to avoid light entering from the side. Oviposition-experienced wasps were individually released at the down-wind end of the Y-tube and observed for a period of 5 min at maximum. If the wasp walked until the end of one of the olfactometer arms, this was recorded as a choice. If the wasp did not reach the end of either arm within 5 min, it was scored as "no response". All wasps were used only once. After every second wasp, the odour sources and the apparatus were exchanged to avoid effects of possible asymmetry.

Bioassay. For the **Y-tube-olfactometer** bioassay two leaves carrying on average 3 egg masses per leaf were excised from an egg-carrying plant and placed with their petiole in a vial with water. The age of the egg masses on these test leaves ranged from 24 to 72 h (see process of egg infestation as described for the contact bioassays). The vial with leaves was transferred to a glass odour container of the olfactometer. This odour source was tested against clean air. About 20 female wasps were tested per leaf pair per day, for a total number of 80 wasps.

Static two-chamber olfactometer tests. To test whether *T. brassicae* wasps are arrested by volatiles from Brussels sprouts plants carrying egg masses of *P. brassicae*, bioassays were conducted in a two-chamber-static-air-flow olfactometer (Figure 1), a slightly modified version of the four-chamber olfactometer described by Steidle and Schoeller (1997). The olfactometer consisted of a cylinder made of acrylic glass (18 cm high, 12 cm ⌀) divided into two chambers by a vertical plate. No airflow

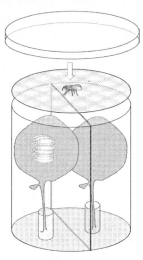

Figure 1: The static 2-chamber olfactometer with two odor fields

was generated. On the top of the cylinder, a removable walking arena (2 cm high, 9 cm ∅) was placed consisting of plastic gauze (mesh 0.1 mm) with a plastic rim and covered with a glass plate. The experiments were carried out in the laboratory at 21 ± 1°C using a fibre optic light source (Euromex coldlight illuminator EK-I, The Netherlands) above the olfactometer. Each oviposition-experienced wasp was individually released in the middle of the two chambers. The time spent by the wasps walking in one of the two odour fields was observed for 300 s. The entire cylinder was randomly rotated after every observation. Each leaf offered was kept with its petiole in a vial with water during the bioassay. Per day and plant 10-15 wasps were tested. In total about 50 wasps were tested with 3-4 plants.

Bioassays. In the static olfactometer bioassay 1, a leaf with 72 h-old egg masses of P. brassicae was placed in one chamber about 5mm below the gauze while the other chamber remained empty. The static olfactometer bioassay 2 was conducted to examine whether T. brassicae is able to differentiate between close-range volatiles from egg-carrying leaves and those from egg-free leaves. Therefore, one chamber was supplied with an egg-carrying leaf (72 h-old eggs), while the other chamber contained a leaf from a "clean" egg-free plant. Both leaves were placed at a distance of about 5mm below the gauze.

Host-age suitability tests. *P. brassicae* eggs of 5 different ages (<12 h, 24 h, 48 h, 72 h and 96 h) were offered on a ca. 1 cm² excised leaf piece from egg-carrying plants (see process of egg infestation as described for the contact bioassays) to 1 day-old mated females of *T. brassicae* with no previous contact to host eggs. An egg clutch consisting of 15 eggs of the same age was offered for a period of 24 h to a female confined in a small glass vial. After that period the wasp was removed from the vial. Eight females were tested per host age and parasitoid species. When the eggs turned black (approximately 5 days after exposure to the parasitoids), the number of parasitized eggs was counted.

Statistics. All contact bioassays and bioassays in the static 2-chamber olfactometer were analysed using Wilcoxon's matched pairs signed rank test. A two-sided binomial test was used to analyse the choices in the Y-tube olfactometer. Parasitism rates for the different host ages were analysed with 5×2 contingency tables and individual χ^2-tests were carried out, corrected by the sequential Bonferroni procedure for table-wide α levels (Rice 1989). The response level of the wasps between the different treatments in contact bioassay 4a) was analysed using Kruskal-Wallis analysis of variance and subsequently compared with Mann-Whitney U-test, corrected by the sequential Bonferroni procedure.

Results

Effects of host eggs and deposits. When leaf squares with eggs and those without eggs were offered, naïve wasps did not discriminate between them. Egg-laden leaf squares were not significantly longer explored by naïve wasps than "clean" egg-free ones (Figure 2, no. 1a, P=0.98, Wilcoxon's matched pairs signed rank test). The naïve wasps spent most of the time in the 'no choice' area (74 %). However, oviposition-experienced wasps clearly preferred leaf squares with eggs deposited 48 h prior to the bioassay (Figure 2, no. 1b, P=0.04, Wilcoxon's matched pairs test). Even when eggs had been removed from the leaf squares, experienced wasps remained significantly longer on such a square when compared to a leaf square which never had received eggs (Figure 2, no. 2, P=0.008, Wilcoxon's matched pairs test). In order to elucidate whether cues from scales and other deposits of *P. brassicae* onto leaves affected *T. brassicae*, we tested leaf squares from plants that had been exposed to the butterflies, but were excluded from egg deposition. Leaf squares with fresh deposits (about 24 h-old) significantly arrested females of *T. brassicae* (Figure 2, no. 3a, P=0.03, Wilcoxon's matched pairs test), whereas leaf squares with 72 h-old deposits did not (Figure 3, no. 3b, P=0.39, Wilcoxon's matched pairs test).

Effect of contact plant cues from egg-carrying plants. In order to examine whether plant surface chemicals in the vicinity of an egg mass serve as cues that indicate the close-by host, leaf squares were excised right next to an egg mass and

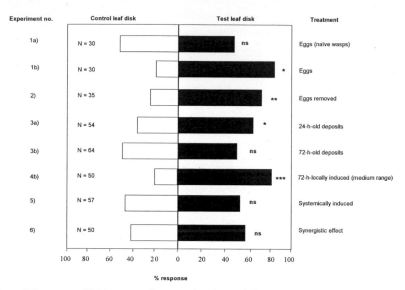

Figure 2: Response of Trichogramma brassicae females on different treated cabbage leaf disks in a two-choice contact bioassay. The percentages of time spent on either the control or test leaf squares is given. Number of tested females is given in Table 1. Asterisks indicate significant differences within the choice test *P<0.05, **P<0.01, ***P<0.001, n.s. not significant (Wilcoxon's matched pairs signed rank test)

offered together with a leaf square from an egg-free "clean" leaf (compare Table 1, experiment no. 4a) The leaf square from a plant, on which eggs had been deposited <12 h prior to the assay significantly arrested *T. brassicae* (Figure 3, P<0.001, Wilcoxon's matched pairs test). The same behaviour was observed when leaf squares from plants, on which eggs had been deposited 24 h prior to the assay were offered. The wasps were significantly arrested on the test square (Figure 3, P<0.001, Wilcoxon's matched pairs test). Females of *T. brassicae* were not arrested on leaf squares cut from an egg-carrying leaf, on which eggs had been deposited 48 h prior to the assay. However, they stayed significantly longer on test leaf squares close to eggs batches deposited 72 h prior to the bioassay (Figure 3, P<0.001, Wilcoxon's matched pairs test). *Trichogramma brassicae* was not only arrested on leaf area in close range of a 72 h-old egg mass, but also on leaf area that was excised about 5 cm away from 72 h-old egg batches (Figure 2, no. 4b, P<0.001, Wilcoxon's matched pairs test). When squares from leaves with 4 day-old eggs were tested, *T. brassicae* still discriminated between the treated and the untreated leaf square (Figure 3, P<0.012, Wilcoxon's matched pairs test). Furthermore, when testing the leaf squares from leaves with about 72 h-old eggs against "clean" squares, females of *T brassicae* spent significantly less time in the "no response" area than they did in the 24 h-old or 48 h-old treatments (Figure 3, P<0.001, Mann Whitney U-test, Bonferroni corrected). Squares taken from egg-free leaves neighbouring egg-laden ones, did not arrest the wasps (Figure 2, no. 5, P=0.598, Wilcoxon's matched pairs test).

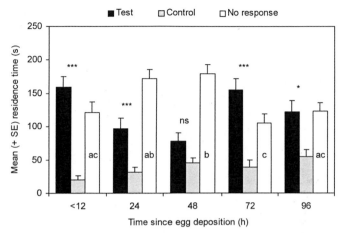

Figure 3: Response of *Trichogramma brassicae* females to "locally oviposition-induced" leaf squares (test) from plants on which eggs had been deposited <12, 24, 48, 72, and 96 hours prior to the bioassay. Compare Table 1, experiment no. 4a. Wasps were tested in a two-choice contact bioassay. Number of tested females per treatment N=50. Mean residence time and standard error are shown. Asterisks indicate significant differences between test and control within the same treatment *P<0.05, ***P<0.001, n.s. not significant (Wilcoxon's matched pairs signed rank test). Different letters indicate significant differences in the no response between the different treatments (Mann-Whitney U-test, Bonferroni corrected)

Effect of larval feeding in combination with egg masses. Leaves carrying eggs for 96 hours and first instar larvae feeding for 48 hours did not arrest females of *T. brassicae* (Figure 2, no. 6, P=0.097, Wilcoxon's matched pairs test).

Effect of volatile plant cues of egg-carrying plants. The wasps' olfactory response to volatiles from leaves carrying eggs for 1-3 days was first tested in a dynamic Y-tube olfactometer. Oviposition-experienced females of *T. brassicae* were not attracted by odour from an egg-carrying leaf, 46% of the 74 responding wasps chose for the egg-carrying leaf (P=0.56, binomial test). When testing the wasps' response to volatiles from egg-carrying leaves at close-range (<5 mm) in a static olfactometer, *T. brassicae* was not arrested by odour from egg-carrying leaves. When offered against clean air, 45% of 48 tested wasps stayed longer above the egg-carrying leaf than above the control leaf (P=0.33, Wilcoxon's matched pairs test). When offered against "clean", egg-free leaves, 49% of 51 tested wasps stayed longer above the egg-carrying leaf than above the control leaf (P=0.80, Wilcoxon's matched pairs test).

Host-age suitability. The highest parasitism rate was found when *T. brassicae* was offered 72 h-old eggs. Host eggs of this age were significantly more frequently parasitized than eggs of any other age tested (Figure 4, 5x2 contingency test, df=4, P<0.001). Eggs that were 96 h old were unsuitable for the wasps showing a low parasitization rate (Figure 4).

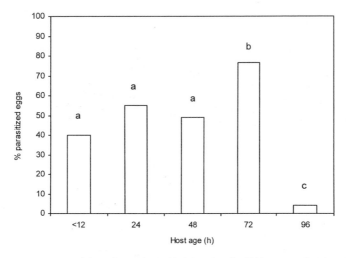

Figure 4: Host-age suitability of eggs from *Pieris brassicae* for *Trichogramma brassicae*. 15 eggs of different ages (<12 h, 24 h, 48 h, 72 h, 96 h) were offered on leaf pieces to 1-day old females for 24 hours. Per egg age a number of 8 females was tested. Different letters in the columns indicate significant differences (P<0.05) (5x2 contingency tables using χ^2).

Discussion

Trichogramma brassicae was arrested on leaf area excised from Brussel sprouts leaves in the vicinity of *P. brassicae* eggs. The factors causing this arrestment change in the course of time after egg deposition has taken place. First, 24 to 48 h after egg deposition, butterfly deposits and eggs on the leaves were found to contribute to this arrestment. This was shown by the wasps' response to egg-free leaves contaminated by butterfly deposits for 24 h and their response to leaves with eggs deposited for 48 h. Second, butterfly deposits were not arresting the wasps when tested 72 h after deposition. Nevertheless, leaf area nearby 72 h-old host eggs had an arresting activity. Eggs of this age were found to be most suitable for parasitisation (Figure 4). From the total dataset the following scenario can be deduced: (a) butterfly deposits and eggs were arresting the parasitoids up to 48 h after oviposition; (b) after 72 h a different effect than of deposits or eggs operates, arresting the wasps in the vicinity of egg masses; (c) the parasitoids did not respond to volatiles from egg-laden leaves, suggesting an induction of arresting leaf surface chemicals perceived after contact; (d) the induction of such leaf surface chemicals seems to be locally restricted since the parasitoids were not arrested on leaf area taken from egg-free leaves that were neighbouring 72 h-old egg-laden ones.

An alternative explanation for possible induction by 72 h-old egg masses could be that substances associated with egg deposition have diffused into the leaf tissue or the leaf's wax layer. Another argument against the induction hypothesis could be that host females deposit a chemical trail around their egg masses. For example, females of *Ostrinia nubilalis* are known to release a secretion around the egg masses by sweeping the abdomen onto the leaf surface during egg deposition. *Trichogramma brassicae* was found to be arrested by this sweeping secretion of *O. nubilalis* (Garnier-Geoffroy et al. 1996). However, trail marking behaviour by females of *Pieris brassicae* has never been observed.

Other studies have shown that oviposition by *P. brassicae* is capable of inducing defences in *Brassica* plants. Shapiro and DeVay (1987) showed that eggs of *Pieris brassicae* and *P. napi* laid on mustard leaves (*Brassica nigra* L. Koch) may induce a hypersensitive response in the plant: 24 hours after egg deposition necrotic zones appear around the eggs leading to a desiccation of the eggs within 3 days. The plant's response was probably elicited by a substance in the glue attaching the eggs on the leaf. Rothschild & Schoonhoven (1977) have demonstrated that *P. brassicae* avoids oviposition on Brussels sprout plants carrying conspecific eggs. This oviposition deterrence was shown to be due to compounds associated with the eggs (Klijnstra 1986). Later studies of Blaakmeer and co-workers showed that none of the oviposition-deterrents, such as miriamides associated with the eggs of *P. brassicae* (Blaakmeer et al. 1994b), were detected anymore on cabbage leaves from which eggs had been removed. Blaakmeer et al. (1994a) suggested that oviposition by *P. brassicae* induces changes at the level of individual leaves resulting in reduced acceptance of these leaves for further oviposition. Even egg-free leaves on a plant laden with eggs during 3-4 days on other leaves were observed to be less accepted for oviposition by *P. brassicae* than leaves of similar age on a plant on which eggs had never been deposited (Blaakmeer et al. 1994a; van Loon, unpubl results), suggesting that changes in plant chemistry rather than cues produced by the ovipositing females or their eggs mediate the effects on conspecific *P. brassicae* females.

While other egg parasitoid species were shown to respond to oviposition-induced plant volatiles (reviewed by Hilker & Meiners, 2002), the *Trichogramma* species tested here did not. Egg deposition by *P. brassicae* on Brussels sprout plants might not have induced volatiles in those quantities and qualities necessary to arrest the egg parasitoids for the particular cultivar studied and under the experimental conditions used. Headspace analyses could reveal whether egg-laden leaves emit volatile blends different from those emitted by egg-free leaves. Also, Noldus & van Lenteren (1983) demonstrated that *T. evanescens* was not attracted by volatiles from cabbage leaves infested with 24 h-old *P. brassicae* eggs and other deposits. Studies about the short distance dispersal ability of *Trichogramma* wasps showed that the major means of this minute organism are walking and short jumps especially at low temperatures (Pak et al. 1985). A directed flight against the wind towards a stimulus source over any significant distance is unlikely (Nordlund 1994). Nevertheless, responses of *Trichogramma* spp. to volatile plant infochemicals have been reported frequently (reviewed by Nordlund 1994; Romeis et al. 1997). However, they were shown to arrest rather than attract the wasps. It is likely that *Trichogramma* spp. use plant cues after entering the habitat. They may reach the habitat/plant passively on wind currents. Another elegant solution to overcome this flight handicap is known for some *Telenomus* spp. These egg-parasitoids explore the sex pheromone of calling moths to locate and mount them and thereafter hitchhike along to the hosts´ egg-laying sites (Arakaki et al., 1995; 1996). Just recently, we were able to show that *Trichogramma brassicae* uses this phoretic strategy as well to reach its hosts eggs by detecting mated *Pieris brassicae* females via an anti-aphrodisiac produced by *Pieris* males (Fatouros et al. 2005). Such a strategy might render the use of oviposition-induced plant volatiles redundant und thus explain the parasitoids´ inability of using induced plant volatiles during habitat location.

The arrestment response by *T. brassicae* to egg-induced leaf surface modifications contrasts with the behaviour of other egg parasitoids which were shown to be attracted to local and systemic oviposition-induced plant volatiles (Colazza et al. 2004b; Hilker et al. 2002; Meiners and Hilker 2000). When comparing the *Brassica-Pieris-Trichogramma* system with the other tritrophic systems studied so far with respect to induction of plant volatiles by insect oviposition, two major differences are obvious (compare Hilker and Meiners 2002 for an overview of these systems).

First, in contrast to specialized egg parasitoids of the other systems investigated so far, *Trichogramma* spp. are known to be fairly polyphagous, parasitizing a variety of mainly lepidopteran host eggs in association with different plant species (Babendreier et al. 2003; Noldus 1989a). According to the concept of dietary specialization and infochemical use in natural enemies it would be expected that they respond to general plant volatiles rather than odours that are specific for a certain plant species (Steidle and van Loon 2003; Vet and Dicke 1992). Indeed, Reddy et al. (2002) were able to show a positive response of *T. chilonis* to two green leaf volatiles, i.e. (Z)-3-hexenyl acetate and hexyl acetate in an olfactometer. Since *T. brassicae* did not respond to any plant volatiles in our olfactometer test, we expect that egg deposition by *P. brassicae* on Brussels sprouts plants does not induce the release of general green leaf volatiles in such amounts that *T. brassicae* becomes attracted. However, learning was shown for *Trichogramma* spp. Rearing the parasitoids on a particular host may be important in this respect (Kaiser et al. 1989). An early adult experience with the rearing host and an oviposition experience can influence preference behaviour of *Trichogramma* wasps (Bjorksten and Hoffmann 1995; Bjorksten and Hoffmann 1998; Kaiser et al. 1989). Rearing *T. brassicae* for several generations in *P. brassicae* and giving them experience with *P. brassicae* eggs on cabbage leaves before each bioassay may have influenced them in their host location behaviour in a way that they start using specific plant cues in close range.

Second, *P. brassicae* females do not damage the plant when laying eggs. The females just briefly drum on the leaf with their forelegs and tap onto the leaf surface with the abdomen to determine its suitability for egg deposition (David and Gardiner 1962). In other systems showing attraction of egg-parasitoids to oviposition-induced plant volatiles, egg deposition was always associated with plant damage: (1) The elm leaf beetle removes the leaf epidermis prior to laying eggs at this site (Meiners and Hilker 2000); (2) The pine sawfly slits a pine needle and lays eggs into such slid needles (Hilker et al. 2002); (3) *Nezara* bugs do not damage bean leaves before egg deposition, but egg parasitoids are only attracted to volatiles from leaves which suffered both egg deposition and feeding damage (Colazza et al. 2004a). Our bioassays showed that even larval feeding associated with egg deposition does not induce volatiles which attract *T. brassicae*.

In conclusion, our results suggest that the response of the generalist egg parasitoids to egg-laden plants differs from the reaction of the specialized wasps studied in other systems. While the latter are attracted by oviposition-induced plant volatiles (Colazza et al. 2004b; Mumm et al. 2003; Wegener et al. 2001), the generalist *T. brassicae* does not respond to volatiles from egg-laden leaves, but instead is arrested by their contact cues. The exact mechanism and the chemistry of these arresting infochemicals need to be elucidated in further studies.

Acknowledgements

The authors thank M.E. Huigens, Torsten Meiners and Joop J.A. van Loon for discussion and comments on an earlier version of this manuscript; Leo Koopman, Frans van Aggelen, and André Gidding for culturing the insects and the experimental farm of the Wageningen University (Unifarm) for breeding and providing the Brussels sprout plants. The study was financially supported by the Deutsche Forschungsgemeinschaft (Hi 416/15-1).

References

Babendreier D, Kuske S, Bigler F (2003) Non-target host acceptance and parasitism by *Trichogramma brassicae* Bezdenko (Hymenoptera: Trichogrammatidae) in the laboratory. Biological Control 26: 128-138

Bjorksten TA, Hoffmann AA (1995) Effects of pre-adult and adult experience on host acceptance in choice and non-choice tests in two strains of *Trichogramma*. Entomologia Experimentalis et Applicata 76:49-58

Bjorksten TA, Hoffmann AA (1998) Persistence of experience effects in the parasitoid *Trichogramma* nr. *brassicae*. Ecological Entomology 23:110-117

Blaakmeer A, Hagenbeek D, van Beek TA, de Groot AE, Schoonhoven LM, van Loon JJA (1994a) Plant response to eggs vs. host marking pheromone as factors inhibiting oviposition by *Pieris brassicae*. Journal of Chemical Ecology 20:1657-1665

Blaakmeer A et al. (1994b) Isolation, identification, and synthesis of miriamides, new hostmarkers from eggs of *Pieris brassicae*. Journal of Natural Products 57:90-99

Boo KS, Yang JP (2000) Kairomones used by *Trichogramma chilonis* to find *Helicoverpa assulta* eggs. Journal of Chemical Ecology 26:359-375

Colazza S, Fucarino A, Peri E, Salerno G, Conti E, Bin F (2004a) Insect oviposition induces volatile emission in herbaceous plants that attracts egg parasitoids. Journal of Experimental Biology 207: 47-53

Colazza S, McElfresh JS, Millar JG (2004b) Identification of volatile synomones, induced by *Nezara viridula* feeding and oviposition on bean spp., that attract the egg parasitoid *Trissolcus basalis*. Journal of Chemical Ecology 30:945-964

David WAL, Gardiner BOC (1962) Oviposition and the hatching of the eggs of *Pieris brassicae* (L.) in a laboratory culture. Bulletin of Entomological Research 53:91-109

Dicke M (1999) Are herbivore-induced plant volatiles reliable indicators of herbivore identity to foraging carnivorous arthropods? Entomologia Experimentalis et Applicata 91:131-142

Fatouros NE, Huigens ME, van Loon JJA, Dicke M, Hilker M (2005) Chemical communication - Butterfly anti-aphrodisiac lures parasitic wasps. Nature 433:704

Garnier-Geoffroy F, Robert P, Hawlitzky N, Frerot B (1996) Oviposition behaviour in *Ostrinia nubilalis* (Lep.: Pyralidae) and consequences on host location and oviposition in *Trichogramma brassicae* (Hym.: Trichogrammatidae). Entomophaga 41:287-299

Hilker M, Kobs C, Varama M, Schrank K (2002) Insect egg deposition induces *Pinus sylvestris* to attract egg parasitoids. Journal of Experimental Biology 205:455-461

Hilker M, Meiners T (2002) Induction of plant responses towards oviposition and feeding of herbivorous arthropods: a comparison. Entomologia Experimentalis et Applicata 104:181-192

Kaiser L, Pham-Delegue MH, Masson C (1989) Behavioural study of plasticity in host preferences of *Trichogramma maidis* (Hym.: Trichogrammatidae). Physiological Entomology 14:53-60

Klijnstra JW (1986) The effect of an oviposition deterring pheromone on egglaying in *Pieris brassicae*. Entomologia Experimentalis et Applicata 41:139-146

Lewis WJ, Martin WRJ (1990) Semiochemicals for use with parasitoids: status and future. Journal of Chemical Ecology 16:3067-3089

Meiners T, Hilker M (2000) Induction of plant synomones by oviposition of a phytophagous insect. Journal of Chemical Ecology 26:221-232

Mumm R, Schrank K, Wegener R, Schulz S, Hilker M (2003) Chemical analysis of volatiles emitted by *Pinus sylvestris* after induction by insect oviposition. Journal of Chemical Ecology 29:1235-1252

Noldus LPJJ (1989a) Chemical espionage by parasitic wasps. How *Trichogramma* species exploit moth sex pheromone systems. In. Agricultural University, Wageningen, p 252

Noldus LPJJ (1989b) Semiochemicals, foraging behaviour and quality of entomophagous insects for biological control. Journal of Applied Entomology 108:425-451

Noldus LPJJ, van Lenteren JC (1983) Kairomonal effects on searching for eggs of *Pieris brassicae*, *Pieris rapae* and *Mamestra brassicae* of the parasite *Trichogramma evanescens* WESTWOOD. Mededelingen van de Faculteit Landbouwwetenschappen Rijksuniversiteit Gent 48:183-194

Nordlund DA (1994) Habitat location by *Trichogramma*. In: Wajnberg E, Hassan SA (eds) Biological Control with Egg Parasitoids. CAB International, Wallingford, Oxon, UK, pp 155-163

Nordlund DA, Lewis WJ, Altieri MA (1988) Influences of plant-induced allelochemicals on the host/prey selection behavior of entomophagous insects. In: Barbosa P, Letournneau DK (eds) Novel aspects of insect-plant interactions. Wiley, New York, USA, pp 65-90

Pak GA, Halder van I, Lindeboom R, Stroet JJG (1985) Ovarian egg supply, female age and plant spacing as factors influencing searching activity in the egg parasite *Trichogramma sp.* Mededelingen van de Faculteit Landbouwwetenschappen Rijksuniversiteit Gent 50:369-378

Reddy GVP, Holopainen JK, Guerrero A (2002) Olfactory responses of *Plutella xylostella* natural enemies to host pheromone, larval frass, and green leaf cabbage volatiles. Journal of Chemical Ecology 28:131-143

Rice WR (1989) Analyzing tables of statistical tests. Evolution 43:223-225

Romeis J, Shanower TG, Zebitz CPW (1997) Volatile plant infochemicals mediate plant preference of *Trichogramma chilonis*. Journal of Chemical Ecology 23:2455-2465

Rothschild M, Schoonhoven LM (1977) Assessment of egg load by *Pieris brassicae* (Lepidoptera: Pieridae). Nature 266:353-355

Schmidt JM (1994) Host recognition and acceptance by *Trichogramma*. In: Wajnberg E, Hassan SA (eds) Biological control with egg parasitoids. CAB International, Wallingford, Oxon, UK, pp 165-200

Schöller M, Prozell S (2002) Response of *Trichogramma evanescens* to the main sex pheromone component of *Ephestia* spp. and *Plodia interpunctella*, (Z,E)-9,12-tetra-decadenyl acetate (ZETA). Journal of Stored Products Research 38:177-184

Shapiro AM, DeVay JE (1987) Hypersensitivity reaction of *Brassicae nigra* L. (Cruciferae) kills eggs of *Pieris* butterflies (Lepidoptera: Pieridae). Oecologia 71:631-632

Steidle JLM, Schoeller M (1997) Olfactory host location and learning in the granary weevil parasitoid *Lariophagus distinguendus* (Hymenoptera: Pteromalidae). Journal of Insect Behavior 10:331-342

Steidle JLM, van Loon JJA (2003) Dietary specialization and infochemical use in carnivorous arthropods: testing a concept. Entomologia Experimentalis et Applicata 108:133-148

Takabayashi J, Dicke M (1992) Response of predatory mites with different rearing histories to volatiles of uninfested plants. Entomologia Experimentalis et Applicata 64:187-193

Vet LEM, Dicke M (1992) Ecology of infochemical use by natural enemies in a tritrophic context. Annual Review of Entomology 37:141-172

Vinson SB (1976) Host selection by insect parasitoids. Annual Review of Entomology 21:109-133

Wajnberg E, Hassan SA (1994) Biological Control with Egg Parasitoids. CAB International, Wallingford, Oxon, UK

Wegener R, Schulz S, Meiners T, Hadwich K, Hilker M (2001) Analysis of volatiles induced by oviposition of elm leaf beetle *Xanthogaleruca luteola* on *Ulmus minor*. Journal of Chemical Ecology 27:499-515

5 Induced Defense Mechanisms in Brussels Sprouts Plants in Response to *Pieris* Egg Deposition: Chemical and Gene Expression Analysis

N.E. Fatouros, S.-J. Zheng, F. Müller, M. Burow, M. Dicke & M. Hilker

Chapter 5

Induced Defense Mechanisms in Brussels Sprouts Plants in Response to *Pieris* Egg Deposition: Chemical and Gene Expression Analysis

Abstract

Plants are known to respond to herbivorous insects by induced defense mechanisms such as emission of volatiles that attract natural enemies. Here, we examined the response of Brussels sprouts plants to egg depositions of the large cabbage white butterfly *Pieris brassicae*. Chemical analysis of leaf surface extracts by gas chromatography – mass spectrometry (GC-MS) and whole leaf contents of glucosinolates by high performance liquid chromatography (HPLC) as well as gene expression analysis of genes encoding the myrosinase (*MYR*) and lipoxygenase (*LOX*) gene were conducted. We compared plants infested with *P. brassicae* eggs (test) that arrest the egg parasitoid *Trichogramma brassicae*, with uninfested plants (control) that do not arrest the parasitoid. No significant differences in epicuticular waxes of leaf surfaces between egg-infested plants and control plants were found. However, a significant decrease of indole-3-acetonitrile was demonstrated in leaf surface extracts of egg-laden plants. Furthermore, no differences in the total amount of all glucosinolates combined in whole-leaf extracts were detected. However, the amount of indolyl-glucosinolates was significantly higher in test than in control plants at 1-4 days after infestation. The *MYR*-gene was shown to be more strongly expressed in plants 1-3 days after egg deposition than in control plants. No induction of expression of the *LOX*-gene was found. Our results are discussed with respect to the response of the egg parasitoid to egg-carrying Brussels sprouts leaves.

Keywords: induced defense, glucosinolates, epicuticular waxes, egg parasitoids, *Pieris*, *Trichogramma*

Introduction

Plants have evolved different strategies to defend themselves against herbivory. By *direct* defense plants negatively affect the herbivore (Karban and Baldwin 1997), and *indirectly* they can defend themselves by luring natural enemies of the herbivores (Dicke 1999). Direct defenses may prevent herbivores from feeding through physical barriers, e.g. thorns or trichomes or through chemical compounds such as secondary plant metabolites or specialized defense proteins. Numerous secondary plant metabolites, including alkaloids and terpenoids, are toxic compounds to herbivores and pathogens and can directly defend plants (Schoonhoven et al. 2005). Volatiles released in response to herbivore feeding guide predators or parasitoids to their herbivorous prey or hosts, and thus benefit both plant and carnivore in such tritrophic systems (Dicke 1999; van Loon et al. 2000). However, herbivore feeding damage to the plant is not required for induction of indirect defense mechanisms. Plants are able to defend themselves indirectly against insect eggs by luring egg parasitoids just before insect feeding starts (Hilker and Meiners 2002; 2006). In three different tritrophic systems it was shown that the plants release volatiles both locally and systemically, induced by egg-deposition in combination with wounding or feeding damage (Colazza et al. 2004a; 2004b; Hilker et al. 2002; Meiners and Hilker 1997; 2000).

In an additional tritrophic system consisting of *Brassica oleracea* var. *gemmifera* (Brussels sprouts plants), *Pieris brassicae* (large white cabbage butterfly), and the egg parasitoid *Trichogramma brassicae* it was shown that leaf area right next to egg depositions by the butterfly arrested the egg parasitoid. One day after egg deposition, this arrestment is due to components deposited by the butterfly along with egg clutches. Two days after egg deposition, these components have lost their arrestment activity. However, after three days, the leaf area in the vicinity to the eggs significantly arrested the egg parasitoids again. A series of bioassays strongly indicated that a change of leaf surface components induced by *P. brassicae* egg deposition after three days arrested the egg parasitoid (Fatouros et al. 2005a).

This study aimed to elucidate the mechanisms of induction in Brussels sprouts plants by butterfly egg deposition both at the chemical and molecular level. Brussels sprouts belong to brassicaceous (cruciferous) plants, which are phytochemically characterized by a class of secondary defense compounds called glucosinolates, a group of thioglucosides with different side-chains. When plant tissue is damaged, glucosinolates are degraded into toxic products, isothiocyanates, thiocyanates, oxazolidine-2-thiones, nitriles, epithionitriles, of which especially the isothiocyanates are known to be toxic for numerous insect species (Renwick 2002;

Teuscher and Lindequist 1994). The degradation process is catalyzed by the enzyme myrosinase (b-thioglucosidase), which is stored in separate cell compartments (Louda and Mole 1991). The glucosinolate-myrosinase system is also referred to as "the mustard oil bomb" (Matile 1980). The glucosinolates and their toxic breakdown products can have diverse biological effects like deterring generalist herbivores or attraction to plants of specialist herbivores of Brassicaceae (reviewed by Rask et al. 2000; Renwick 2002). Nevertheless, few studies claimed that some of the breakdown products negatively affect specialist herbivores of Brassicaceae (Agrawal and Kurashige 2003; Li et al. 2000).

Brussels sprouts plants were also shown to emit a blend of volatiles when infested by *P. brassicae* or *P. rapae* caterpillars, and these blends were shown to attract larval parasitoids such as *Cotesia* spp. (Agelopoulos and Keller 1994; Geervliet et al. 1994; Mattiacci et al. 1995). Headspace collections from *Pieris*-infested plants revealed a quantitative increase of several compounds (Blaakmeer et al. 1994; Mattiacci et al. 1995) and 20 of them evoked electroantennogram responses in *Cotesia* parasitoids (Smid et al. 2002).

In this study, the effect of egg deposition on Brussels sprout plants was studied with respect to the following questions regarding chemical and molecular mechanisms:

(a) Does egg deposition by *P. brassicae* affect the amounts of glucosinolates in egg-laden leaves?

(b) Does egg deposition induce a change in the chemical composition of cuticular waxes on the surface of egg-laden leaves?

(c) Are the possible oviposition-induced changes of the glucosinate contents due to an enhanced transcription of myrosinase?

(d) Does oviposition induce an increase of transcription of the *LOX* gene such as is known for induction by feeding?

It is known from the behavioral studies of the effects of egg-laden Brussels sprouts leaves on *T. brassicae* that the time of induction is important: only after three days since egg deposition, the induced arrestment effect becomes evident. Thus, to follow the induction process in the course of time, the questions b, c, and d were studied on a time scale ranging from one day to five days.

Materials and methods

Plants and insects. *Brassica oleracea* L. var. *gemmifera* cv. Cyrus plants, 8-12 weeks old, were reared in a greenhouse (18 ± 2°C, 70% rh, L16:D8). *Pieris brassicae* L. (Lepidoptera: Pieridae) was reared on Brussels sprouts plants in a climate room (21 ± 1°C, 50-70% r.h., L16:D8). Each day a plant was placed into a cage (80 x 100 x 80 cm) with ca. 100 adult butterflies for approximately 24 h to allow egg deposition. The emerging larvae were reared on Brussels sprouts plants until adulthood.

Plant treatments. An 8-to-10-week-old plant was introduced into a cage with ca. 100 butterflies to allow egg-deposition for 8 h (local), and two leaves were covered with gauze (ca. 1 mm^2 mesh size) to avoid egg-deposition (systemic). Control plants had not been in contact with the butterflies. Plants were kept in a climate room (21 ± 1°C, 50-70% r.h., L16: D8). A treated plant was on average laden with approximately six egg masses per leaf.

For leaf surface analysis, two leaves of each treatment (local and systemic) and of the uninfested plant were excised 72 h after egg deposition. To obtain an extract the surface of these leaves was treated as described below.

For glucosinolate analysis leaf material (100 - 500 mg fresh weight) was collected 24, 96, and 120 h after egg deposition from local and systemic leaves as well as from uninfested plants from respective sites. Egg hatching began 120 h after egg deposition but larvae did not feed yet. After excision, the leaf parts were immediately frozen in liquid nitrogen, and thereafter freeze-dried over night.

For gene-expression analysis, leaf material of local and systemic leaves was collected in the same way at 24, 72, 96, and 120 h after egg deposition and immediately placed in liquid nitrogen.

Extraction of leaf surface compounds. Before extraction egg-batches from egg-carrying plants were removed with a fine spatula. The leaves of each treatment were successively dipped about 6 cm deep into dichloromethane (DCM) (Roth, Germany) for 3 s to remove the non-polar waxes, subsequently dipped for a 5 s period into methanol (MeOH) (Roth, Germany) to retrieve more polar substances. Extracts were then kept at -20° till analysis.

GC/MS analysis of leaf surface extracts. Extracts were analyzed with coupled gas chromatography and mass spectrometry (GC-MS) on a Fisons GC model 8000 and a Fisons MD 800 quadrupole MS using a J&W 30 m DB5-ms capillary column

(0.32 mm internal diameter, film thickness 0.25 μm). Per sample, 1 μl was injected in splitless mode (injector temperature 240 °C) with helium as the carrier gas (inlet pressure 10 kPa). The temperature program started at 40 °C (4 min hold) and was raised with 10 °C/min up to 280 °C (20 min hold). EI (70 eV) mass spectra were recorded.

Both underivatized and derivatized samples (N = 8 - 9) were analyzed. DCM-fractions were derivatized by TMSH (0.2M trimethylsulfoniumhydroxid in MeOH). Samples (1μl) were co-injected with 1μl TMSH, the derivatization reaction took place at 280 °C injector temperature. Components were identified by comparing mass spectra and linear retention indices with NIST library spectra or with those of authentic reference compounds. Components were quantified by the quotient of the peak areas compound and internal standard.

Glucosinolate analysis. For quantification of glucosinolates, 10 - 30 mg freeze-dried leaf material was ground to a fine powder in a paint shaker (Kliebenstein et al. 2001) and extracted with 1.5 ml 80 % (v/v) methanol for 1 min at room temperature after addition of 50 μl 1 mM 4-hydroxybenzylglucosinolate as internal standard. Unsoluble material was pelleted by centrifugation at $2.500 \times g$ for 10 min and the supernatants were loaded on columns prepared with 0.4 ml of a 10 % (w/v) suspension of DEAE Sephadex A25 in water. Columns were washed with 1 ml 80 % (v/v) methanol, 1 ml water, and 1 ml 0.02 M MES buffer, pH 5.2, before 50 μl of purified sulfatase solution was applied (Hogge et al. 1988). After incubation at room temperature overnight, desulfated glucosinolates were eluted with 2 x 0.8 ml 60 % (v/v) methanol, dried at 50 °C in a nitrogen stream, and then redissolved in 0.4 ml water. Samples were analyzed by HPLC on an Agilent HP1100 Series instrument equipped with a C-18 reverse phase column (LiChrospher RP18ec, 250 x 4.6 mm, 5 μm particle size) (compare Kliebenstein et al., 2001) using a water (solvent A)-acetonitrile (solvent B) gradient at a flow rate of 1 ml min-1 at 25°C (injection volume 50 μl). The gradient was as follows: 1.5 % B (1 min), 1.5 - 5 % B (5 min), 5 - 7 % B (2 min), 7 - 21 % B (10 min), 21 - 29 % B (5 min), 29 - 43 % B (7 min), 43 - 100 % (0.5 min), 100 % B (2.5 min), 100 - 1.5 % B (0.1 min), and 1.5 % B (4.9 min). The eluent was monitored by diode array detection between 190 and 360 nm (2 nm interval). Desulfoglucosinolates were identified based on comparison of retention times and UV absorption spectra with those of known standards (Reichelt et al. 2002). Results are given as μmol (g dry weight)$^{-1}$ calculated from the peak areas at 229 nm relative to the peak area of the internal standard (relative response factor 2.0 for aliphatic glucosinolates and 0.5 for indole glucosinolates). N = 8-10 extract samples were analyzed from each treatment type and control.

RNA isolation. Total RNA isolation from control and treated *Brassica* leaves was performed according to the method described by Zheng et al. (2005) with minor modifications. Briefly, 0.75 ml of RNA extraction buffer (100 mM Tris pH 8.5, 100 mM NaCl, 20 mM EDTA, 1% Sarkosyl) with ß-mercaptoethanol was added to each fine powder sample. After vigorously vortexing, 0.75 ml of buffer-saturated phenol was added, and the sample was vortexed again. Following centrifugation at 14,000 rpm for 15 min, 500 µl phenol/chloroform (1:1) was added to the aqueous phase and centrifuged at 14,000 rpm for 10 min. The samples were left at -80 °C for 15 min and then centrifuged at 14,000 rpm for 10 min at 4 °C. The supernatant was removed and 500 µL of DNase/Rnase-free DEPC-treated water was added. After adding LiCl to a final concentration of 2 M, the samples were incubated overnight at 4 °C. The samples were centrifuged at 14,000 rpm for 10 min at 4 °C and the supernatant was discarded. The RNA pellet was dissolved completely by adding 400 µl of DNase/Rnase-free DEPC-treated water. After dilution, 40 µl of 3M NaAc (pH 5.2) and 1 ml ethanol (96%) were added. The samples were kept at -80 °C and after 10 min spun down at 14,000 rpm for 5 min at 4 °C. The supernatant was discarded, and the RNA pellet was left to dry on air for 5 min. The dried RNA pellet was dissolved by adding 50 µl of DNase/Rnase-free DEPC-treated water. The concentration of RNA was measured with a BIO-RAD SmartSpec™3000.

cDNA cloning of cabbage myrosinase and LOX-2. First-strand cDNA was synthesized using 2 µl oligo-dT primer, 2 µl 2′-deoxynucleoside S′-triphosphates (dNTPs), 8 µl first-strand buffer, 4 µl 0.1 M DTT, 2 µl RNase OUT and 2 µl M-MLV reverse transcriptase (Invitrogen, the Netherlands) and 5 µg RNA from *P. rapae*-infested leaves in a total volume of 40 µl. The sample was incubated at 25 °C for 10 min and 37 °C for 50 min. The reaction was stopped by incubating at 70 °C for 15 min. Degenerated primers for myrosinase were designed from conserved regions of *Arabidopsis thaliana* (NCBI gene accession NM122499, NM122501 and X89413); *Brassica napus* (NCBI gene accession X60214 and X82577) and *Brassica juncea* (NCBI gene accession AJ223494). The forward primer used was 5′-5′GCRTGGYCAAGAMTMHTTCC3′-3′ and the reverse primer used was 5′-GARCAWCGWMCNGGTGC -3′. PCR was performed in a total of 50 µl containing 0.2 µM primers, 0.2 mM dNTPs, 0.4 units Super Taq DNA polymerase (Invitrogen), 1 X PCR buffer (Invitrogen), and 2 µl of first-strand cDNA. Temperature programs were: 2 min at 94 °C; 40 cycles of 1 min at 94 °C, 1 min at 45 °C, 1 min at 72 °C; and 10 min at 72 °C. The PCR products were separated using agarose gel electrophoresis. Bands were cut from the gel and purified using the QIAEX II Gel extraction kit (Qiagen) according to the manufacturer's instructions. Aliquots of

3 µL purified PCR products were ligated into the vector pGEM® -T (Promega, Wisconsin, USA) according to the instructions of the manufacturer. After overnight ligation at 4 °C, supercompetent cells from *Escherichia coli* Epicurian Coli® XL2-Blue MRF' (Stratagene, California, USA) were transformed with the ligation mixture by a heat shock as suggested by the supplier. Ampicillin–resistant colonies were identified and plasmid DNA was isolated using a miniprep kit (Qiagen) according to the manufacturer's instructions. Plasmids were selected which showed inserts of the expected size after digestion with *Nco*I and *Sal*I. Expected inserts based on size were used for sequencing. Sequencing was performed in an ABI PRISM 310 automated DNA sequencer (Perkin Elmer, USA). Cabbage myrosinase accession number was obtained from NCBI. Cloning of the *LOX-2* has been described by Zheng et al. (2006).

Gene expression analysis. Isolated RNA was DNAase treated to remove genomic DNA contamination in RNA samples. For the cDNA synthesis, 5 mg of total RNA was used. First-strand cDNA was synthesized as mentioned above. The PCR program used for gene expression was 2 min at 94°C; 5 cycles of 30 sec at 94°C, 3 min at 72°C; 5 cycles of 30 sec at 94°C, 30 sec at 70°C, 3 min at 72°C; 13 cycles of 30 sec at 94°C, 30 sec at 68°C, 3 min at 72°C. The gene specific primers MYF1 (5'AG GGGAGTGAACCAAGGAGGTCTTGA'3) and MYR1 (5'CGGGTGCATCTGTTC CGATAGCATAG'3) produced 280 bp products. Touchdown PCR was also used for the housekeeping gene *GAPDH* as a loading control (Zheng et al. 2006). N = 3-5 samples of each treatment type and control were used to determine expression rates of the *MYR* and *LOX-2* gene.

Statistics. For the chemical analysis of leaf surface extracts and glucosinolate levels, samples of treated and control leaves were collected on the same day and therefore analyzed as paired data. The Wilcoxon's matched-pair signed rank test was employed to test whether the quotient of peak area/ internal standard per compound in the leaf surface extracts differed between control and local leaves, or local and systemic leaves (Compare Fatouros et al. 2005b). The same test was used to assess whether the amount of glucosinolates differed between control and local leaves or control and systemic leaves.

Results and discussion

Five predominant glucosinolates were found in egg-laden and untreated leaves of Brussels sprouts plants: three aliphatic (3-methylsulfinyl-propylglucosinolate [trivial name: glucoiberin], 4-methylsulfinyl-butylglucosinolate [glucoraphanin] and 2-propenyl-allylglucosinolate [sinigrin]) as well as two indole glucosinolates (indol-3-ylmethylglucosinolate [glucobrassicin] and 4-methoxy indol-3-ylmethylglucosinolate [4-methoxyglucobrassicin]). Except for the glucoraphanin, these glucosinolates have been already identified before in whole leaf extracts of *Brassica oleracea* plants (Renwick et al. 1992).

We compared the quantities of glucosinolates of leaves from untreated plants (control) with those from egg-laden (local) and egg-free leaves (systemic) of plants on which eggs had been deposited 24, 96, and 120 h before. When considering the total amount of these five glucosinolates, no differences were found between the control and the treated plants, nor between egg-laden and egg-free leaves of the same plant and treatment (Figure 1). However, regarding the quantities of specific

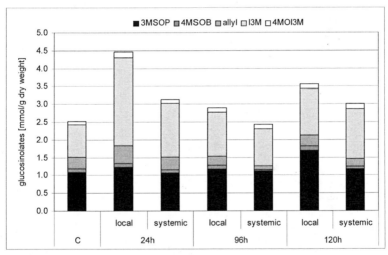

Figure 1: Amounts of glucosinolates in Brussels sprouts plants infested with *P. brassicae* eggs 24, 96, and 120 h after egg deposition vs. uninfested plants (C). Detected aliphatic glucosinolates (dark bars): 3-methylsulfinyl-propyl (3MSOP), 4-methylsulfinyl-butyl (4MSOP) and 2-propenyl-allyl (allyl). Detected indole glucosinolates (light bars): indol-3-yl methyl (I3M) and 4-methoxyindol-3-ylmethyl (4MOI3M).

glucosinolates, some significant differences were detected. When considering egg-laden leaves 24 h after egg-deposition, a significant increase was found in the amount of indole glucosinolates compared to leaves of untreated plants (Figure 1, P=0.007, Wilcoxon's matched pairs signed rank test). After 96 h, the indole glucosinolates were still significantly increased in egg-infested local leaves (Figure1, P=0.033, Wilcoxon's matched pairs signed rank test). No systemic effect of egg deposition was found, neither when comparing glucosinolate quantities with control leaves nor with local leaves. The increase of indole glucosinolates in egg-laden local leaves gives an indication for an induced response of Brussels sprouts plants to *P. brassicae* egg deposition after 24 up to 96 h.

The effect of glucosinolates on specialist herbivores is contradictory. Glucosinolates were demonstrated to act as feeding and oviposition stimulants for specialist lepidopterous insects such as *Pieris* spp. (reviewed by Rask et al. 2000). Glucobrassicin, an indole glucosinolate, was shown to stimulate the oviposition in *Pieris rapae* and *P. brassicae* (Renwick et al. 1992; van Loon et al. 1992). By using an enzyme that prevents isothiocyanates from being formed after ingestion or an enzyme that redirects glucosinolate hydrolysis toward the less toxic nitrile formation, the *Brassica* specialists *Plutella xylostella* and *P. rapae* are able to 'disarm the mustard oil bomb' (Ratzka et al. 2002; Wittstock et al. 2004). In contrast, some studies claim that previously damaged plants negatively affected *P. rapae* by induced responses of mustard plants (Agrawal 2000; Agrawal and Kurashige 2003; Traw and Dawson 2002), but no direct link with glucosinolates was established. In contrast, generalist herbivores have generally been seen to be deterred by increased levels of glucosinolates and exhibited reduced growth and survival (reviewed by Rask et al., 2000).

A plant's response to egg deposition in terms of increasing the amounts of glucosinolates already before larval feeding starts could have two benefits: a) other (generalist) herbivores could be deterred from oviposition and b) hatching larvae could suffer from reduced performance after ingesting the glucosinolate breakdown products. However, whether the increase that we recorded is strong enough to reduce the performance of specialists remains unclear.

Figure 2 shows the expression of the myrosinase gene (*MYR*), which was detected in most control plants as well as in egg-infested plant treatments. Gene expression was considered to be induced by egg-infestation if at least three of the five replicates from egg-infested plants showed higher expression than the replicates from undamaged control plants. Expression of the *MYR*-gene was induced in local egg-laden leaves after 24 and 72 h and in systemic, egg-free leaves of egg-infested plants after 72 h (Figure 2). The number of replicates for the 120h treatment was too low to draw a reliable conclusion from. Thus, we indicated that myrosinase transcription was induced by *P. brassicae* egg deposition, first only locally in egg-laden leaves and after 3 days also systemically. Some replicates showed double bands probably due to DNA rests.

Recent studies indicated that the enzyme myrosinase might also have a role in defense against herbivores in addition to catalyzing the glucosinolate hydrolysis (Li et al. 2000; Mitchell-Olds et al. 1996). Higher myrosinase activity in *Brassica* plants reduced feeding of *P. xylostella* larvae. Because myrosinase activity is related

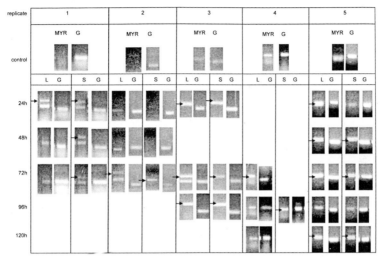

Figure 2: Induction of myrosinase (*MYR*) gene expression in Brussels sprouts plants after *P. brassicae* egg deposition. RNA was extracted from undamaged plants (control) or plants where eggs had been deposited 24, 48, 72, 96, or 120 h before extraction. RNA was extracted locally from egg-infested leaves (L) and from systemic leaves (S) of the same plant. *GAPDH* (G) was analyzed as the housekeeping gene, a constitutive control. Arrows indicate an induction of the *MYR* gene expression.

to isothiocyanate concentrations it was suggested that at sufficient concentrations of myrosinase the resulting isothiocyanates could be deterrent to feeding crucifer specialists (Li et al. 2000). Yet, this relation does not apply to indole glucosinolates whose breakdown products do not result in the toxic isothiocyanates. Instead, glucobrassicin typically degrades by enzymatic breakdown into the non-volatiles indol-3-acetonitrile, indole-3-carbinol and 3,3-diindolylmethane (Chevolleau et al. 1997 and references therein). Indol-3-ylmethanol (I3M) can polymerize into even less volatile oligomers (Fenwick et al. 1983). The ecological impact of the increase of *MYR*-gene transcription and indole glucosinolates on *P. brassicae* remains to be tested.

Leaf surface extracts of whole leaves were made from egg-laden plants onto which eggs had been deposited 72 h before and analyzed with GC-MS. Egg-laden leaves (local) were compared with egg-free leaves (systemic) from the same plant and with uninfested control plants. The leaf washes made with dichloromethane revealed four compounds typical for epicuticular waxes: two long-chained alkanes, a long-chain fatty acid, and an alkanediol (Figure 3). The relative abundance of the two alkanes and the fatty acid was significantly different in locally induced leaves compared to systemic leaves from the same plant (Figure 3, P<0.05, Wilcoxon's matched pairs signed rank test), but did not differ from those in the control plants. This could be an indication for a local induction by the egg clutches of the three mentioned compounds within individual plants. When comparing different plant individuals such small chemical changes might be overruled by genetic variation. However, in bioassays with *T. brassicae*, the wasps were able to discriminate between leaf disks from egg-laden plants and uninfested plants (Fatouros et al. 2005a).

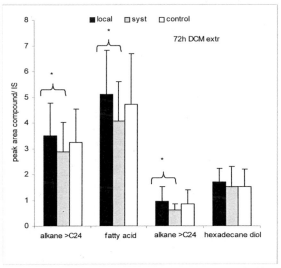

Figure 3: Compounds detected through GC-MS analysis in DCM extracts of cabbage leaf surfaces. Extracts were made from local, egg-infested (grey) and systemic, egg-free leaves (dark) from the same plant 72h after treatment, as well as from leaves of untreated plant (white). Mean values of the quotient peak area compound / peak area internal standard (C13 + C24) (IS) ± standard deviation are shown. Wilcoxons' matched pairs signed rank test, * p<0.05. Number of analyzed plants per treatment N=9.

Long-chain alkanes from cuticular host products like egg shells or moth scales were shown to arrest *Trichogramma* spp. (Jones et al. 1973; Renou et al. 1992; Shu et al. 1990). Cuticular lipids in plants are structurally very similar or even identical to those of insects (Espelie et al. 1991). It was shown that the behavior of herbivorous insects is influenced by plant cuticular lipids. Specific surface components can enhance or deter herbivory or oviposition (Espelie et al. 1991; Müller and Riederer 2005). Some studies also describe the influence of plant cuticular waxes in interactions with parasitoids. An increased amount of the triterpene squalene in the apple leaf surface induced by leaf miners was shown to serve as host location cue for the braconid wasp *Pholetesor bicolor* (Dutton et al. 2002). Another braconid, *Cotesia glomerata*, was arrested on the edge of *P. rapae* -damaged leaves and to some long-chain fatty acids involved that originated from the plant surface. It was assumed that the fatty acids are produced locally along the host-fed edge as an interaction between the plant and the larval regurgitant (Horikoshi et al. 1997).

After derivatization, additional compounds were detected in the DCM leaf surface extracts (Figure 4). The alkene tricosene was identified at slightly higher levels on the leaf surface of egg-laden leaves. This compound is known as a sex pheromone component of hymenopteran, dipteran and lepidopteran species (Carlson et al. 1971; El-Sayed 2005; Gibb et al. 2006).

In methanol extracts from leaf surfaces we found three different compounds: a sulfide, benzyl cyanide, and a glucobrassicin derivative namely indole-3-acetonitrile. The glucobrassicin derivative was found in significantly higher

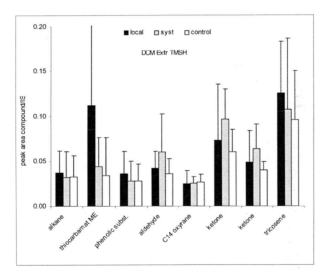

Figure 4: Compounds detected through GC-MS in DCM extracts of cabbage leaf surfaces after derivatization with TMSH. Extracts were made from local, egg-infested (grey) and systemic, egg-free leaves (dark) from the same plant 72h after treatment, as well as from leaves of untreated plants (white). Mean values of the quotient peak area compound / peak area internal standard (C13 + C24) (IS) ± standard deviation are shown. Number of analyzed plants per treatment N=8.

amounts on leaf surfaces of untreated plants than on surfaces from systemic leaves of egg-laden plants (Figure 5, P<0.01, Wilcoxon's matched pairs signed rank test) and in tendency in reduced amounts on leaf surfaces of egg-laden leaves. This decrease of the glucobrassicin derivate in leaf surface extracts is opposite of what we found in whole leaf extracts: here the amount of glucobrassicin was increased in egg-laden plants 24 h after egg deposition. It remains unclear if the glucosinolates are actually present on the leaf surface or escape from the inner leaf. Reifenrath

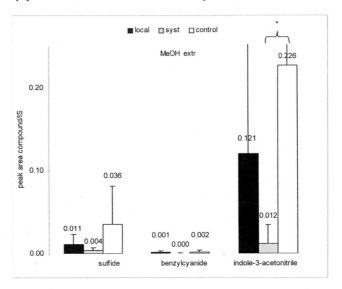

Figure 5: Compounds detected through GC-MS in MeOH extracts of cabbage leaf surfaces. Extracts were made from local, egg-infested (grey) and systemic, egg-free leaves (dark) from the same plant 72h after treatment, as well from leaves of untreated plant (white). Mean values of the quotient peak area compound / peak area internal standard (C12 OH) (IS) ± standard deviation are shown. Wilcoxons' matched pairs signed rank test, * p<0.05. Number of analyzed plants per treatment N=9.

et al. (2005) did not detect the polar glucosinolates on the outer wax layer when mechanically removing it whereas extraction led to varying glucosinolate profiles depending on the light conditions the plants were kept. The authors concluded that the glucosinolates might escape from the inner leaf tissue through the open stomata and are therefore found in solvent extracts, instead of being naturally present on the outer wax layer.

In our system, we expected a role of the plant cuticular substances as host finding cues for *T. brassicae*. Except for the decrease in a glucobrassicin derivate, we did not detect any obvious changes in the chemical composition of the leaf surface after egg deposition of *P. brassicae*. This could also be due to the fact that the compounds arresting *T. brassicae* on leaf area right next to an egg mass three days after egg deposition were insoluble in the solvents we used. Only hot boiling water released the oviposition stimulating chemicals for *P. rapae* from the leaf surface of cabbage plants (Städler and Roessingh 1990). Yet, van Loon et al. (1992) could stimulate oviposition of *P. brassicae* with a methanol-post-dichloromethane extract.

Furthermore, because it is difficult to identify the chemosensory detection used by the *T. brassicae* wasps in our bioassays we can not exclude that short-range plant volatiles induced by *P. brassicae* egg deposition were involved. Egg deposition by the elm leaf beetle (*Xanthogaleruca luteola*) and the pine sawfly (*Diprion pini*) were shown to induce volatile emissions in the elm tree and pine tree consisting mainly of terpenoids and attracting specialist egg parasitoids (Hilker et al. 2002; Meiners and Hilker 2000; Mumm et al. 2003; Wegener and Schulz 2002; Wegener et al. 2001). However, unlike in the Brussels sprouts - *Pieris* system, both herbivorous insects damage the leaf prior to egg deposition, which might cause a different response in the two tree systems by releasing long-range plant volatiles.

Analysis of the *LOX-2*-gene expression after egg-deposition did not show support for the involvement of the octadecanoid pathway in the plants response to insect eggs after 1-4 days (data not shown). The plant hormone jasmonic acid (JA) is produced by the octadecanoid pathway and plays a central role in signal transduction in insect-inducible indirect defense (van Poecke and Dicke 2004). Upregulation of *LOX-2*-gene expression was demonstrated in the same *Brassica oleracea* cultivars as used here after larval feeding by *P. brassicae* (Zheng et al. 2006).

Conclusions

Secondary metabolites present in the epicuticular wax layer or compounds of the wax layer itself were shown to play an important role in insect-plant interactions (Müller and Riederer 2005). Our leaf surface analysis of Brussels sprouts plants gave a hint that the amount of the glucobrassicin breakdown product indole-3-acetonitrile decreases three days after oviposition by *P. brassicae*. Furthermore, we found evidence for an increase in myrosinase activity and indolyl-glucosinolate contents in total leaves.

We hypothesize that a change in the glucosinolates by egg deposition in Brussels sprouts plants could play a role in indirect defense by arresting *T. brassicae* wasps.

Since the specialist *Pieris* spp. are known to use glucosinolates in the leaf surface as stimulants for oviposition or feeding (Rask et al. 2000; Renwick 2002), these compounds could also act as indirect cues in host location by natural enemies such as *Trichogramma* egg parasitoids. Indeed, volatile isothiocyanates were shown to attract the aphid parasitoid *Diaeretiella rapae* to *Brassica* plants (Bradburne and Mithen 2000). Yet, plants induced by feeding caterpillars of *P. brassicae* were not arresting *T. brassicae* wasps (Fatouros et al. 2005a), although it is known that herbivore feeding leads to a breakdown of the glucosinolates and to the toxic isothiocyanates. However, as far as *Trichogramma* wasps and *Pieris* butterflies are concerned, only chemical "information" available at the surface is relevant. Induction of glucosinolates due to feeding might lead to different chemical profiles than induction by eggs, as it was indicated here. To test this hypothesis further chemical and behavioral studies are needed.

Acknowledgements

The authors thank Joop J. A. van Loon for discussion and comments on an earlier version of this manuscript, Leo Koopman, Frans van Aggelen, and André Gidding for culturing insects, and the experimental farm of the Wageningen University (Unifarm) for breeding and providing the Brussels sprouts plants. The study was financially supported by the Deutsche Forschungsgemeinschaft (Hi 416/15-1/2).

References

Agelopoulos NG, Keller MA (1994) Plant-natural enemy association in the tritrophic system, *Cotesia rubecula-Pieris rapae*- Brassicaceae (Cruciferae).III: Collection and identification of plant and frass volatiles. Journal of Chemical Ecology 20:1955-1967

Agrawal AA (2000) Benefits and costs of induced plant defense for *Lepidium virginicum* (Brassicaceae). Ecology 81:1804-1813

Agrawal AA, Kurashige NS (2003) A role for isothiocyanates in plant resistance against the specialist herbivore *Pieris rapae*. Journal of Chemical Ecology 29:1403-1415

Blaakmeer A, Geervliet JBF, van Loon JJA, Posthumus MA, van Beek TA, de Groot Æ (1994) Comparative headspace analysis of cabbage plants damaged by two species of *Pieris* caterpillars: consequences for in-flight host location by *Cotesia* parasitoids. Entomologia Experimentalis et Applicata 73:175-182

Bradburne RP, Mithen R (2000) Glucosinolate genetics and the attraction of the aphid parasitoid *Diaeretiella rapae* to *Brassica*. Proceedings of the Royal Society of London Series B-Biological Sciences 267:89-95

Carlson DA, Mayer MS, Silhacek DL, James JD, Beroza M, Bierl BA (1971) Sex attractant pheromone of the house fly: isolation, identification and synthesis. Science 174:76-78

Chevolleau S, Gasc N, Rollin P, Tulliez J (1997) Enzymatic, Chemical, and Thermal Breakdown of ^{3}H-Labeled Glucobrassicin, the Parent Indole Glucosinolate. J. Agric. Food Chem. 45:4290-4296

Colazza S, Fucarino A, Peri E, Salerno G, Conti E, Bin F (2004a) Insect oviposition induces volatile emission in herbaceous plants that attracts egg parasitoids. Journal of Experimental Biology

207:47-53

Colazza S, McElfresh JS, Millar JG (2004b) Identification of volatile synomones, induced by *Nezara viridula* feeding and oviposition on bean spp., that attract the egg parasitoid *Trissolcus basalis*. Journal of Chemical Ecology 30:945-964

Dicke M (1999) Evolution of induced indirect defense of plants. In: Tollrian R, Harvell CD (eds) The Ecology and Evolution of Inducible Defences. Princeton University Press, Princeton, NJ, pp 62-88

Dutton A, Mattiacci L, Amadò R, Dorn S (2002) A novel function of the triterpene squalene in a tritrophic system. Journal of Chemical Ecology 28:103-116

El-Sayed AM (2005) The Pherobase: Database of Insect pheromones and Semiochemicals In. Ashraf M. El-Sayed

Espelie KE, Bernays EA, Brown JJ (1991) Plant and insect cuticular lipids serve as behavioral cues for insects. Archives of Insect Biochemistry and Physiology 17:223-233

Fatouros NE, Bukovinszkine'Kiss G, Kalkers LA, Soler Gamborena R, Dicke M, Hilker M (2005a) Oviposition-induced plant cues: do they arrest *Trichogramma* wasps during host location? Entomologia Experimentalis et Applicata 115:207-215

Fatouros NE, Van Loon JJA, Hordijk KA, Smid HM, Dicke M (2005b) Herbivore-induced plant volatiles mediate in-flight host discrimination by parasitoids. Journal of Chemical Ecology 31:2033-2047

Fenwick GR, Heany RK, Mullin WJ (1983) Glucosinolates and their breakdown products infood and food plants. CRC Critical Reviews in Food Science and Nutriotion 18:123-201

Geervliet JBF, Vet LEM, Dicke M (1994) Volatiles from damaged plants as major cues in long-rage host-searching by the specialist parasitoid *Cotesia rubescula*. Entomologia Experimentalis et Applicata 73:289-297

Gibb AR, Suckling DM, Morris BD, Dawson TE, Bunn B, Dymock JJ (2006) (Z)-7-tricosene and monounsaturated ketones as sex pheromone components of the Australian guava moth *Coscinoptycha improbana*: identification, field trapping, and phenology. Journal of Chemical Ecology 32:221-237

Hilker M, Kobs C, Varama M, Schrank K (2002) Insect egg deposition induces *Pinus sylvestris* to attract egg parasitoids. Journal of Experimental Biology 205:455-461

Hilker M, Meiners T (2002) Induction of plant responses towards oviposition and feeding of herbivorous arthropods: a comparison. Entomologia Experimentalis et Applicata 104:181-192

Hilker M, Meiners T (2006) Early herbivore alert: insect eggs induce plant defense. Journal of Chemical Ecology:in press

Hogge LR, Reed DW, Underhill EW, Haughn GW (1988) Hplc separation of glucosinolates from leaves and seeds of *Arabidopsis thaliana* and their identification using thermospray liquid-chromatography mass-spectrometry. Journal of Chromatographic Science 26:551-556

Horikoshi M, Takabayashi J, Yano S, Yamaoka R, Ohsaki N, Sato Y (1997) *Cotesia glomerata* female wasps use fatty acids from plant-herbivore complex in host searching. Journal of Chemical Ecology 23:1505-1515

Jones RL, Lewis WJ, Beroza M, Bierl BA, Sparks AN (1973) Host-seeking stimulants (kairomones) for the egg parasite, *Trichogramma evanescens*. Environmental Entomology 2:593-596

Karban R, Baldwin IT (1997) Induced Responses to Herbivory. The Univerity of Chicago Press, Chicago

Kliebenstein DJ, Lambrix VM, Reichelt M, Gershenzon J, Mitchell Olds T (2001) Gene duplication in the diversification of secondary metabolism: tandem 2-oxoglutarate-dependent dioxygenases control glucosinolate biosynthesis in *Arabidopsis*. Plant Cell 13:681-693

Li Q, Eigenbrode SD, Stringham GR, Thiagarajah MR (2000) Feeding and growth of *Plutella xylostella* and *Spodoptera eridania* on *Brassica juncea* with varying glucosinolate concentrations and myrosinase activities. Journal of Chemical Ecology 26:2401-2419

Louda S, Mole S (1991) Glucosinolates: Chemistry and ecology. In: Rosenthal GA, Berenbaum M (eds) Herbivores: Their Interactions with Secondary Plant Metabolites, vol I. The Chemical Participants, 2nd edn. Academic Press, San Diego, California, pp 123-164

Matile P (1980) The mustard oil bomb - compartmentation of the myrosinase system. Biochemie und Physiologie der Pflanzen 175:722-731

Mattiacci L, Dicke M, Posthumus MA (1995) ß-Glucosidase: An elicitor of the herbivore-induced plant odor that attracts host-searching parasitic wasps. Proceedings of the National Academy of Sciences of the United States of America 92:2036-2040

Meiners T, Hilker M (1997) Host location in *Oomyzus gallerucae* (Hymenoptera: Eulophidae), an egg parasitoid of the elm leaf beetle *Xanthogaleruca luteola* (Coleoptera: Chrysomelidae). Oecologia 112:87-93

Meiners T, Hilker M (2000) Induction of plant synomones by oviposition of a phytophagous insect. Journal of Chemical Ecology 26:221-232

Mitchell-Olds T, Siemens D, Pedersen D (1996) Physiology and costs of resistance to herbivory and disease in *Brassica*. Entomologia Experimentalis et Applicata 80:231-237

Müller C, Riederer M (2005) Plant surface properties in chemical ecology. Journal of Chemical Ecology 31:2621-2651

Mumm R, Schrank K, Wegener R, Schulz S, Hilker M (2003) Chemical analysis of volatiles emitted by *Pinus sylvestris* after induction by insect oviposition. Journal of Chemical Ecology 29: 1235-1252

Rask L, Andreasson E, Ekbom B, Eriksson S, Pontoppidan B, Meijer J (2000) Myrosinase: gene family evolution and herbivore defense in Brassicaceae. Plant Molecular Biology 42:93-113

Ratzka A, Vogel H, Kliebenstein DJ, Mitchell-Olds T, Kroymann J (2002) Disarming the mustard oil bomb. Proceedings of the National Academy of Sciences of the United States of America 99: 11223-11228

Reichelt M et al. (2002) Benzoic acid glucosinolate esters and other glucosinolates from *Arabidopsis thaliana*. Phytochemistry 59:663-671

Reifenrath K, Riederer M, Müller C (2005) Leaf surface wax layers of Brassicaceae lack feeding stimulants for *Phaedon cochleariae*. Entomologia Experimentalis et Applicata 115:41-50

Renou M, Nagan P, Berthier A, Durier C (1992) Identification of compounds from the eggs of *Ostrinia nubilalis* and *Mamestra brassicae* having kairomone activity on *Trichogramma brassicae*. Entomologia Experimentalis et Applicata 63:291-303

Renwick JAA (2002) The chemical world of crucivores: lures, treats and traps. Entomologia Experimentalis et Applicata 104:35-42

Renwick JAA, Radke CD, Saqchdev-Gupta K, Städler E (1992) Leaf surface chemical stimulating oviposition in *Pieris rapae* (Lepidoptera: Pieridae). Chemoecology 3:33-38

Schoonhoven LM, van Loon JJA, Dicke M (2005) Insect-Plant Biology, 2nd edn. Oxford University Press, Oxford

Shu S, Swedenborg PD, Jones RL (1990) A kairomone for *Trichogramma nubilale* (Hymenoptera: Trichogrammatidae). Isolation, identification, and synthesis. Journal of Chemical Ecology 16:521-529

Smid HM, Van Loon JJA, Posthumus M, A., Vet LEM (2002) GC-EAG-analysis of volatiles from Brussels sprouts plants damaged by two species of *Pieris* caterpillars: olfactory respective range of a specialist and generalist parasitoid wasp species. Chemoecology 12:169-176

Städler E, Roessingh P (1990) Perception of surface chemicals by feeding and ovipositing insects. In: Szentesi A, Jermy T (eds) 7th International Symposium on Insect-Plant Relationships (SIP), vol. 39. Symp Biol Hung, Hungary, pp 71-86

Teuscher E, Lindequist U (1994) Biogene Gifte. Biologie, Chemie, Pharmakologie, 2 edn. Gustav Fischer Verlag

Traw MB, Dawson TE (2002) Reduced performance of two specialist herbivores (Lepidoptera: Pieridae, Coleoptera: Chrysomelidae) on new leaves of damaged black mustard plants. Enviromental Entomolgy 31:714-722

van Loon JJA, Blaakmeer A, Griepink FC, Van Beek TA, Schoonhoven LM, De Groot AE (1992) Leaf surface compound from *Brassica oleracea* (Cruciferae) induces oviposition by *Pieris brassicae*. Chemoecology 3:1-6

van Loon JJA, de Boer JG, Dicke M (2000) Parasitoid-plant mutualism: parasitoid attack of herbivore increases plant reproduction. Entomologia Experimentalis et Applicata 97:219-227

van Poecke RMP, Dicke M (2004) Indirect defence of plants against herbivores: Using *Arabidopsis thaliana* as a model plant. Plant Biology 6:387-401

Wegener R, Schulz S (2002) Identification and synthesis of homoterpenoids emitted from elm leaves after elicitation by beetle eggs. Tetrahedron 58:315-319

Wegener R, Schulz S, Meiners T, Hadwich K, Hilker M (2001) Analysis of volatiles induced by oviposition of elm leaf beetle *Xanthogaleruca luteola* on *Ulmus minor*. Journal of Chemical Ecology 27:499-515

Wittstock U et al. (2004) Successful herbivore attack due to metabolic diversion of a plant chemical defense. Proceedings of the National Academy of Sciences of the United States of America 101:4859-4864

Zheng SJ, Bruinsma M, Dicke M (2006) Sensitivity and speed of induced defense of cabbage (*Brassica oleracea* L.): dynamics of *BoLOX* expression patterns during insect and pathogen attack.

Zheng SJ, Henken B, de Maagd RA, Purwito A, Krens FA, Kik C (2005) Two different *Bacillus thuringiensis* toxin genes confer resistance to beet armyworm (*Spodoptera exigua* Hubner) in transgenic Bt-shallots (*Allium cepa* L.). Transgenic Research 14:261-272

6 | Butterfly Anti-Aphrodisiac Involved in Eliciting Oviposition-Induced Plant Responses

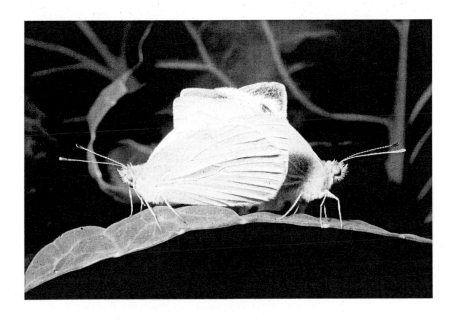

N.E. Fatouros, G. Bukovinszkine'Kiss, J.J.A. van Loon, M.E. Huigens, M. Dicke & M. Hilker

Chapter 6

Butterfly Anti-Aphrodisiac Involved in Eliciting Oviposition-Induced Plant Responses

Abstract

Egg deposition by herbivorous insects can induce chemical plant defense responses that "recruit" egg parasitoids. The egg parasitoid *Trichogramma brassicae* is known to be arrested by chemical changes in the leaf surface of Brussels sprouts plants that are induced by egg deposition of the large cabbage white butterfly *Pieris brassicae* after 3 days. Here, we investigated the elicitor inducing these leaf surface changes. Eliciting activity was found in a secretion associated with the eggs, i.e., secretion of the accessory reproductive glands (ARG) of mated female *P. brassicae*. In contrast, ARG secretion obtained from virgin females had no activity. Male butterflies are known to transfer an anti-aphrodisiac to females during mating to repel competing males. During egg deposition the female butterfly excretes this pheromone, benzyl cyanide, together with the ARG secretion onto the leaf surface of the host plants. When benzyl cyanide alone was applied in large amount (20 µg) onto the leaf surface, an eliciting activity was detected, whereas a 10-times lower amount had no eliciting activity anymore. However, a mixture of the anti-aphrodisiac close to the natural amount detected in ARG secretion from mated females (1 ng), together with the ARG secretion from virgin females induced leaf surface changes that significantly arrested the wasps 3 days after application. Thus, the male anti-aphrodisiac transferred to the females activates the female ARG secretion so that it can act as elicitor of plant defense induced by egg deposition. Recently, benzyl cyanide was shown to play a role for foraging *T. brassicae* in another context, i.e., by acting as a cue to engage in phoretic transport by mated female butterflies to egg-laying sites. Our results indicate that these two independent mechanisms triggered by an anti-aphrodisiac pheromone incur fitness costs to the butterflies.

Keywords: egg parasitoid, elicitor, induction, tritrophic interactions, infochemical, *Pieris*, *Trichogramma*

Introduction

Chemical signals play a crucial role in the interactions between herbivorous insects and parasitic wasps (Hilker and Meiners 2002a; Stowe et al. 1995; Vinson 1984). In order to locate tiny eggs of herbivorous host insects in an ocean of plant biomass, egg parasitoids have been shown to employ chemical cues a) from the plant induced by host egg-deposition (Fatouros et al. 2005a; Hilker and Meiners 2002b) or b) from the adult host stage (i.e. infochemical detour) (Vet and Dicke 1992) rather than cues emanating from the eggs themselves.

Plants under herbivore attack often start to release chemical cues attracting predators and parasitoids that effectively defend the plant by killing the herbivores (reviewed by Dicke 1999; Hilker and Meiners 2002b). For some tritrophic systems it was shown that this indirect plant defense is triggered by compounds in the herbivore's regurgitant allowing the plant to discriminate between artificial wounding and insect feeding (Alborn et al. 1997; Halitschke et al. 2001; Mattiacci et al. 1995).

However, plants can also respond prior to being damaged by feeding insects (reviewed by Hilker and Meiners 2002b; 2006; Hilker et al. 2002b). Egg deposition by phytophagous insects has been shown to induce volatiles attractive to egg parasitoids in tritrophic interactions associated with elm (Meiners and Hilker 1997; 2000), pine (Hilker et al. 2002a; Mumm et al. 2003), and bean plants (Colazza et al. 2004a; b). The elicitor of these oviposition-induced synomones was shown to be located in the oviduct secretion of the female herbivore (Hilker et al. 2002a; b; Meiners and Hilker 2000). The oviduct secretion of herbivorous insects is used to envelop and/or glue the eggs onto the leaves. It is produced by accessory reproductive glands (ARG) that secrete into the reproductive tract (Gillott 2002).

Earlier investigations showed that products associated with the eggs may also have a kairomonal effect on egg parasitoids. Noldus and van Lenteren (Noldus and van Lenteren 1985) demonstrated an arresting effect of fresh extracts of Pieris brassicae (Lepidoptera: Pieridae) egg washes to Trichogramma evanescens Bezdenko (Hymenoptera: Trichogrammatidae). Furthermore, it was shown for other egg parasitoids that kairomones from the host accessory reproductive gland (ARG) secretion played a role during host recognition (Nordlund et al. 1987; Strand and Vinson 1982; 1983).

Recent data showed that egg deposition by the large cabbage white butterfly *P. brassicae* L. on Brussels sprouts plants (*Brassica oleracea* var. *gemmifera* cv.

Cyrus) induces chemical changes of the leaf surface next to the eggs that arrest the generalist egg parasitoid *T. brassicae* (Fatouros et al. 2005a). These changes of the leaf surface were not induced immediately after egg deposition, but became apparent locally at the egg-laden leaf only after 3 days when the host eggs are most suitable for parasitisation. Thus, 3 days after egg deposition leaf parts of the egg-laden leaf that had not been into contact with eggs arrested *T. brassicae* wasps (Fatouros et al. 2005a).

The aim of this study was to investigate which egg-associated components are involved in induction of the leaf surface changes used by *T. brassicae*. We especially focused on the *P. brassicae* female ARG secretion that is associated with the eggs. Furthermore, we are interested in the possible role of male products transferred to the females during mating in this induction by egg deposition. Males are known to transfer a spermatophore to females that contains sperm, nutrients, and male ARG products during mating (Boggs and Gilbert 1979; Gillott 2002; 2003). The male ARG components exert their effects at all phases of the reproductive biology of the mated female: from the moment sperm is deposited in the reproductive tract to the time egg deposition (Gillott 2002; 2003; Leopold 1976). In butterflies, anti-aphrodisiacs were found to be transferred during copulation. The substances curtail courtship and decrease the likelihood of female remating (Andersson et al. 2000; 2003; Gilbert 1976). In the large cabbage white butterfly *P. brassicae*, males transfer benzyl cyanide (BC), a component of their own body odor and of their spermatophores (Andersson et al. 2003), to females. BC acts as a kairomone by luring the egg parasitoid *T. brassicae* to mated *P. brassicae* females. The wasp subsequently uses the female as a transport vehicle to their oviposition sites (Fatouros et al. 2005b). Here, we hypothesize that this anti-aphrodisiac is also involved in eliciting chemical modifications in the leaf surface of Brussels sprout plants laden with eggs.

Material and methods

Plants and insects. Brussels sprout plants (*Brassica oleracea* L. var. *gemmifera* cv. Cyrus) were reared in a greenhouse ($18 \pm 2°C$, 70% rh, L16:D8). Plants of 8-12 weeks old having ca. 14-16 leaves were used for the rearing of *Pieris brassicae* and for the experiments.

Pieris brassicae L. (Lepidoptera: Pieridae) was reared on Brussels sprouts plants in a climate room ($21 \pm 1°C$, 50-70% r.h., L16:D8). Each day a plant was placed into a large cage (80 x 100 x 80 cm) with more than 100 adults for 24 hours to allow egg deposition. Virgin males and females were obtained by sexing in the pupal phase and subsequently containing them separately in a cage. Mating pairs of *P. brassicae* were isolated to obtain mated females and males. As soon as a pair was observed to finish mating, females and males were isolated and their reproductive tissue dissected.

Trichogramma brassicae Bezdenko (Hymenoptera: Trichogrammatidae) (strain Y175) was reared in eggs of *P. brassicae* for several generations (25 ± 1 °C, 50-70% rh, L16:D8). For the rearing, 1-3 day old *P. brassicae* eggs on leaves were used. Only mated, 2-5 days old, oviposition-experienced female wasps were used for the experiments. An oviposition experience was given for a period of 18 hours prior to the experiment with 1-3 day old *P. brassicae* eggs deposited on Brussels sprout leaves.

Preparation of reproductive tissue. To obtain samples for the bioassays, three gravid or virgin *P. brassicae* females, 5-14 days old, were dissected in PBS (phosphate buffered saline, pH 7.2) and the accessory glands were transferred to a vial with 100 µl PBS and homogenized. Then further 100 µl PBS was added. The homogenate was centrifuged for 5 min at 14.000 rpm, and 100 µl of the supernatant was applied with a brush onto the edge of a cabbage leaf as described below. When not used immediately, the accessory gland homogenate was stored at -80 °C.

For GC-MS analysis of the accessory glands, three glands of either gravid or virgin *P. brassicae* females, 5-14 days old, were transferred to a vial with 50 µl hexane, homogenized and centrifuged for 10 min at 14.000 rpm. A volume of 20 µl of the supernatant was transferred to a microtube for analysis. In total 6 samples of each gland type were analyzed. The reproductive tract of a *P. brassicae* male, 4-14 days old and either recently mated (< 3 hours) or virgin, was dissected in PBS and transferred to a vial containing 30 µl hexane and 1000 ng of methyl oleate as an internal standard was added. The bursa copulatrix of mated females was treated in the same way for the analysis (N= 10 samples).

Bioassay. A wasp was individually released in the centre of a small glass Petri dish (5.5 cm diameter) lined with filter paper. The wasp was simultaneously offered a test and a control leaf square of ca. 2 x 2 cm each. These squares were cut from plants treated as described below. Test and control leaf squares were placed approx. 1 cm apart. A wasp was released in the centre between the two leaf squares. The total duration of time spent on the leaf squares (with antennal drumming during most of the time) was observed for a period of 300 sec using The Observer 3.0 software (Noldus Information Technology 1993©). The time spent searching in the area outside the leaf disks was scored as "no response". A total of 50 wasps per treatment was tested, 10 wasps per experimental day and plant. Leaf squares were renewed after every third wasp tested.

Plant treatments. First bioassays were conducted to test whether the behavior of the egg parasitoid is affected by the ARG homogenate itself. Either 10 µl homogenate of gravid or virgin *P. brassicae* females was applied on a leaf square of an egg-free plant and tested against a leaf square with 10 µl PBS in a two choice bioassay. The bioassay was started 5 min after application of the homogenate and buffer, respectively, onto the test and control leaf square.

In order to test whether female ARG secretion contains components inducing a local plant response arresting the egg parasitoid, an ARG sample was applied onto the edge of a Brussels sprouts leaf. In the middle between the tip and the base of the leaf, a stretch of about 2 cm of the edge was treated on the lower leaf side. After treatment, these test plants were kept for 24, 72, or 96 hours in a climate chamber (21±1°C, 50-70% rh, L16:D8). Control plants were treated in the same way with PBS only. Fatouros et al. (Fatouros et al. 2005a) showed that eggs of *P. brassicae* induce a change in the leaf surface right next to the eggs 72 hours after oviposition, but not yet after 24 hours. Female *T. brassicae* wasps are arrested on this locally induced egg-free leaf area cut from a leaf next to 72-hours-old *P. brassicae* eggs. To test the local effect of the putative elicitor in ARG secretion, test leaf squares near the treated leaf part were tested against leaf squares from the untreated part of the control plant in a two-choice bioassay after 24 or 72 hours.

To test a systemic effect, leaf squares were taken from an untreated leaf (systemic leaf) of an ARG-treated plant, 96 hours after treatment. A leaf right above the treated leaf was used as a systemic leaf. Control leaf squares were cut from a systemic leaf of a plant treated with PBS.

In order to examine whether pure benzyl cyanide induces a local plant response arresting the wasps 24 or 72 hours after application, a volume of 100 µl of 200 ng BC/µl PBS was applied onto the edge of a Brussels sprout leaf as described above for the ARG secretion. Thus, the total amount of BC was 20 µg per leaf. Two other BC concentrations, i.e., 100 ng/µl and 20 ng/µl were tested additionally 72 hours after application. These BC concentrations had also been used in bioassays testing the kairomonal activity on the wasps (Fatouros et al. 2005b). Twenty microgram BC was applied per butterfly when testing the intraspecific activity of the anti-aphrodisiac among the butterflies (Andersson et al. 2003). A control plant was treated with PBS only. Leaf squares from an untreated part of the test leaf were tested against leaf squares from the untreated part of the control leaf in the two-choice bioassay.

To test for a possible systemic effect of BC, leaf squares were taken from an untreated leaf right above the treated leaf (systemic leaf) 96 hours after treatment with BC (100 µl of 200 ng BC/µl PBS). For this experiment, the treated leaf was covered with a plastic bag for 96 hours to prevent that volatiles released from the treated leaves adsorb on the systemic leaf. To maintain gas exchange, an incoming and outgoing airstream (200 ml/min) was supplied that was cleaned by a charcoal filter. The control plant was treated in the same way as described above and control squares were obtained from the respective systemic leaf.

To test whether a low concentration of BC induces a local plant response when combined with ARG from virgin females, either 10 ng or 0.1 ng benzyl cyanide (BC), were diluted in 10 µl hexane and added to 200 µl of the gland homogenate of virgin females in PBS. A volume of 100 µl of this mixture was applied onto a leaf as described above, i.e., 100 ng or 1 ng BC plus gland homogenate equivalents from three females were used per leaf. The test leaf squares were cut from the plant after 72 hours since application of the mixture. Control leaf squares were obtained from leaves treated with PBS and 10 µl hexane. Again, leaf squares from an untreated part of the test leaf were tested against leaf squares from the untreated part of the control leaf in the two-choice bioassay.

Chemical analysis of reproductive tissue. Accessory gland extracts of mated and virgin females were analyzed by coupled gas chromatography – mass spectrometry (GC-MS) on a Fisons GC model 8060 and a Fisons MD 800 quadrupole MS using a J&W 30 m DB5-ms capillary column (0.32 mm internal diameter, film thickness 0.25 µm). Per sample, 1 µl was injected in splitless mode (injector temperature 240 °C) with helium as the carrier gas (inlet pressure 10 kPa). The temperature program started at 40 °C (4 min hold) and was raised with 10 °C/min up to 280 °C. The column effluent was ionized by electron impact ionization (EI) at 70 eV. Mass spectra were recorded with full scan mode from 35-350 m/z with a scan time of 0.9 s and interscan delay of 0.1 s. Identification of the compound BC was made by comparison of retention time and mass spectra with an authentic reference sample of BC (99 %, Aldrich). The amounts of BC were quantified in single ion mode (SIM).

The reproductive tract of males and the bursa copulatrix of females were also analyzed by gas chromatography (GC). As internal standard, 1000 ng of methyl oleate was added to the samples. Each sample was sealed and stored for about 24 hours at -20°C before analysis. Five microliter of each sample were injected into a GC 8000 *TOP SERIES* (Carlo Erba Instruments) equipped with a 30 m x 0.25 mm DB-5 column (thickness 0.25 μm) and a flame ionization detector. The oven temperature was programmed from 80-200°C at 10°C/min followed by 200-255°C at 3°C/min. BC and methyl oleate were identified by comparison with reference compounds obtained from Pherobank, Wageningen, The Netherlands. Peak areas of both compounds were compared in the program GC-EAD version 2.6 (Syntech NL) to calculate the amount of BC.

Statistics. All bioassays were analyzed using Wilcoxon's matched pairs signed rank test. The amount of BC in the reproductive tissue was analyzed using a Mann-Whitney-U signed rank test.

Results

When *T. brassicae* wasps could contact ARG homogenate of mated *P. brassicae* females applied onto leaf squares, they were significantly arrested (P=0.04, Wilcoxon's matched pairs signed rank test, figure 1a) five minutes following application, whereas fresh ARG homogenate obtained from virgin females did not elicit a response (P=0.76, Wilcoxon's matched pairs signed rank test, Figure 1a).

When untreated leaf squares were cut from leaf surface close to leaf area treated with female ARG homogenate, the wasps' responses to such untreated squares were the following: Neither treatment by ARG homogenate of mated females (P=0.62, Wilcoxon's matched pairs signed rank test, figure 1b) nor by ARG homogenate of virgin females (P=0.99, Wilcoxon's matched pairs signed rank test, figure 1b) arrested the wasps when tested 24 hours after treatment. However, 72 hours after ARG application, the wasps spent significantly more time on untreated leaf squares taken from sites adjacent to area treated with ARG homogenate of mated females than on control leaf squares from plants treated with the solvent PBS only (P=0.004, Wilcoxon's matched pairs signed rank test, Figure 1c). Thus, these arresting leaf squares are denoted here as `locally induced' ones. In contrast, ARG homogenate of virgin *P. brassicae* females applied 72 hours before the behavioral test did not elicit such an effect (P=0.86, Wilcoxon's matched pairs signed rank test, figure 1c).

A systemic effect of ARG homogenate was found as well after 96 hours of induction. The wasps spent significantly more time on leaf squares taken from an entirely untreated leaf of a plant treated with ARG homogenate obtained from mated females than on control leaf squares (P=0.02, Wilcoxon's matched pairs

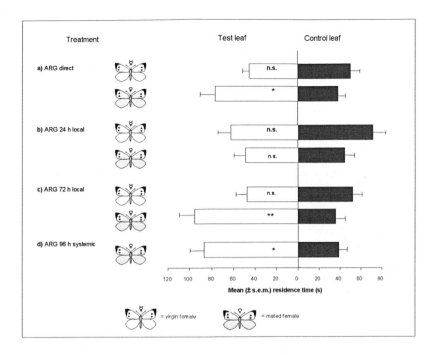

Figure 1: Arrestment of *Trichogramma brassicae* on leaves treated with accessory reproductive gland (ARG) homogenate of either virgin or mated *Pieris brassicae* females (test), all diluted in PBS versus leaves treated with PBS only (control). ARG direct: leaf area treated with ARG homogenate was offered to the parasitoid. ARG 24 h (72 h) local: untreated leaf area, but adjacent to a site on the same leaf that was treated 24 h (72 h) previously with ARG homogenate was offered to the parasitoid. ARG 96 h systemic: untreated leaf area from an untreated leaf that is adjacent to a leaf treated 96 h before with ARG homogenate was offered. Wasps were tested in a two-choice bioassay. Number of tested females per treatment: N=50. Mean residence time and standard error are shown. Asterisks indicate significant differences between test and control within the same treatment. *P<0.05, **P<0.01, n.s. not significant (Wilcoxon's matched pairs signed rank test).

signed rank test, Figure 1b).

When untreated leaf squares were cut from leaf surface close to leaf area treated with BC instead of ARG homogenate, the wasps' responses to these locally induced squares were the following: They were not arrested by leaf squares cut from leaves on which 20 µg BC was applied 24 hours before the test (P=0.36 Wilcoxon's matched pairs signed rank test, Figure 2a). However, after 72 hours the wasps discriminated leaf squares plants treated with the same amount of BC from leaf squares of PBS-control plants (P=0.04, Wilcoxon's matched pairs signed rank test, Figure 2b). Yet, the two lower concentrations did not induce any plant response eliciting a behavioral effect in the egg parasitoid.

A systemic effect of BC was found 96 hours after leaf treatment. Leaf squares obtained from an untreated systemic leaf located above a leaf treated with BC (200 ng/µl PBS) significantly arrested *T. brassicae* wasps compared to leaf squares from an PBS-treated plant (*P*=0.001, Wilcoxon's matched pairs signed rank test, Figure 2c).

Locally induced leaf squares from leaves treated with ARG homogenate of virgin females to which 100 ng BC was added, arrested the wasps after 72 h of induction (*P*=0.007, Wilcoxon's matched pairs signed rank test, Figure 2d). A similar response was recorded to locally induced leaf squares from plants treated with ARG homogenate of virgins in combination with 1 ng BC (*P*=0.04, Wilcoxon's matched pairs signed rank test, Figure 2d).

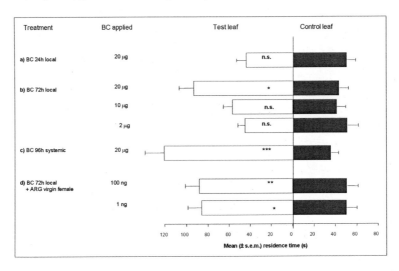

Figure 2: Arrestment of *Trichogramma brassicae* to leaves treated with benzyl cyanide (BC) (test) or leaves treated with ARG homogenate of virgin *P. brassicae* and BC, all diluted in PBS versus leaves treated with PBS only (control). BC 24 and 72 h local: untreated leaf area, but adjacent to a site on the same leaf that was treated 24 h (72 h) previously with BC was offered to the parasitoid. BC 96 h systemic: untreated leaf area from an untreated leaf that is adjacent to a leaf treated 96 h before with BC was offered. ARG and BC 72 h local: untreated leaf area, but adjacent to a site on the same leaf that was treated 72 h previously with ARG homogenate of virgin *P. brassicae* combined with BC was offered to the parasitoid. Wasps were tested in a two-choice bioassay. Number of tested females per treatment N=50. Mean residence time and standard error are shown. Asterisks indicate significant differences between test and control within the same treatment. *P<0.05, **P<0.01, ***P<0.001, n.s. not significant (Wilcoxon's matched pairs signed rank test).

GC-MS analysis showed traces of benzyl cyanide (BC) (up to 1 ng) in all hexane-extracts of accessory reproductive glands of mated females of *Pieris brassicae* (Figure 2, Table 1). No BC traces were found in the ARG-extracts of virgin females (Figure 2). The reproductive tract of virgin males contained on average 2449 ± 749 ng BC (Table 1). The amount of BC was significantly lower in mated males (591 ± 171 ng BC, $P=0.006$, Mann-Whitney-U-Test, Table 2). When analyzing the amount of BC in the bursa copulatrix of mated females, in 9 of 10 samples no BC was detected, while a single female contained 85 ng BC in her bursa (Table 1).

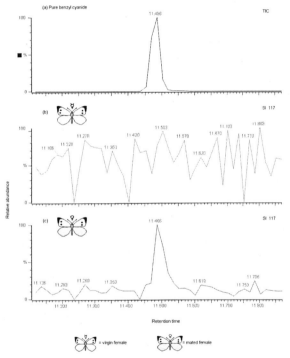

Figure 3: GC-MS analyses of (a) pure benzyl cyanide, (b) accessory reproductive gland (ARG) extracts of virgin *Pieris brassicae* females, or (c) ARG extracts of mated *Pieris brassicae* females. Number of tested samples per treatment N=6. Chromatogram (a) shows a part of the total ion mode (TIC). BC was only found in traces in the glands of mated females, therefore the relative abundance of the only main mass 117 (SI=single ion) is shown the chromatogram (b) and (c). Numbers in the graph are retention times (min).

Discussion

Our results show that application of ARG secretion of mated females induced a local plant response arresting the egg parasitoid only after 72 hours, but not after 24 hours. Accordingly, also egg depositions induced this plant response after 72 hours (Fatouros et al. 2005a). In contrast, ARG secretion of virgin females did not induce such a plant response, suggesting that mating changes the composition of the ARG secretion. Our results show that the male anti-aphrodisiac BC that is transferred to the female during mating plays a role for the eliciting activity of ARG secretion of

Table 1: Amount of benzyl cyanide (ng) in the reproductive tissue of male and female *P. brassicae*

Tissue	Gender	Mating status	BC in ng (mean ± se)	N	P-value
Reproductive tract	Male	virgin	2449 ± 749	5	0.006
		mated	591 ± 171	8	
Accessory gland	Female	virgin	0	6	0.001
		mated	± 1	6	
Bursa copulatrix	Female	mated	85	1	

mated females.

When testing BC in such quantities as they occur in the ARG secretion of mated females in combination with ARG secretion from virgin females, this mixture also elicited a plant response arresting the egg parasitoid locally 72h after application. BC *per se* did not act as elicitor in concentrations close to those in the ARG secretion of mated females. Our results suggest that BC "activates" the eliciting function of the ARG secretion of virgin females, since this secretion gains eliciting activity when BC is added. Thus, a single component that the female received from the male during mating changes the activity of the ARG secretion in such a way that defensive plant responses to egg deposition are induced.

Few studies so far have addressed the chemistry of the elicitor of oviposition-induced plant responses. A small protein of 10 – 12 kDa has been suggested to act as elicitor of an oviposition-induced plant response in another tritrophic system consisting of the Scots pine (*Pinus sylvestris*), the eggs of the pine sawfly (*Diprion pini*), and its egg parasitoid *Chrysonotomyia ruforum*. The oviduct secretion of the sawfly contains this candidate substance that elicits after application onto pine needles the release of pine volatiles attracting the egg parasitoid (Hilker et al. 2005). Oviduct secretion of the elm leaf beetle *Xanthogaleruca luteola* is also known to elicit a plant response in elm leaves that results in the release of volatiles attracting egg parasitoids (Meiners and Hilker 2000). However, the chemical character of this elicitor is unknown so far. The chemistry of the elicitor of a direct plant defensive response to egg deposition is known only from bruchid beetles. When bruchid beetles lay eggs onto pea pods of a specific genotype, neoplasms are formed that elevate the egg from the pea pod surface. The neoplasm hinders the larva to enter the pod, or enhances the chance that eggs drop off the plant and fall to the ground. The elicitors are long-chain α, γ-monounsaturated C_{22}-diols and α, γ-mono- and diunsaturated C_{24}-diols, mono- or diesterified with 3-hydroxypropanoic acid (39).

The known elicitors of feeding induced defensive plant responses are: (a) fatty-acid-amino-acid conjugates (Alborn et al. 2000; 1997; Halitschke et al. 2001; Pohnert et al. 1999), including the well-known volicitin (Turlings et al. 2000), and (b) β-glucosidase from *Pieris brassicae* oral secretions (Mattiacci et al. 1995). When one of these components is applied onto plant wounds, they elicit the release of volatiles that repel ovipositing herbivores (De Moraes et al. 2001), or attract natural enemies of herbivores (Dicke and van Loon 2000). The exact plant perception mechanisms of elicitors from herbivores are poorly known. A study on volicitin in the regurgitant of *S. exigua* showed that the initiation of the maize plant's response to feeding damage was mediated by a protein-ligand interaction in plasma membrane fractions (Truitt et al. 2004). Elicitors isolated from pathogens and their interactions with the plant surface have been more intensively. Here, receptor-mediated activation of ion channels or direct interaction with lipid bilayers leading to pore formation appear to be possible mechanisms of elicitor action (Klüsener and Weiler 1999).

Wounding of the plant tissue is considered a condition for the functioning of the elicitor in the oviduct secretion of the pine sawfly and the elm leaf beetle (Hilker et al. 2005; Meiners and Hilker 2000). While pine sawflies severely damage the pine needle with the slerotized ovipositor valves and applies the oviduct secretion deeply into the needle's wound, the elm leaf beetle just nibbles upon the leaf epidermis prior to egg deposition at this nibbled site (Hilker and Meiners 2006). Except of some slight scratches of the wax layer, no wounding of the leaf tissue is observed when *P. brassicae* lays its eggs. Thus, the female ARG secretion needs to contain components that enable the elicitor to pass through the waxy plant cuticula to the plant cells which can initiate a response to egg deposition.

The anti-aphrodisiac BC is not the only component transferred from male *P. brassicae* to the females. In *Pieris* spp., the male ejaculate has at least three effects: fertilizing the eggs of the male's mate, reducing the receptive period of the females, and increasing her fecundity and longevity through nutrients (Bissoondath and Wiklund 1996; Wiklund et al. 1993). We cannot exclude that these additional components in the male ejaculate also play a role for the eliciting activity of mated female ARG secretion. BC as well as these additional components might also induce changes of the female ARG secretion that are important for the eliciting activity of

ARG secretion of mated females.

Compounds found in the oviduct secretions of *P. brassicae*, the so-called miriamides, were first believed to constitute an oviposition-deterring pheromone preventing *P. brassicae* from egg deposition on (Blaakmeer et al. 1994b). It was Blaakmeer et al. (1994a) who proposed that oviposition deterrence in *P. brassicae* in response to egg-infested cabbage leaves was not caused by the miriamides, but by the host plant itself producing an allomone in response to egg deposition. It was suggested that the miriamides function differently by protecting the plant against predators or fungal infections (Blaakmeer et al. 1994a). It cannot be ruled out that miriamides are responsible for a leaf surface modulating effect in Brussels sprouts after egg deposition. The eliciting activity of a mixture of synthetic BC and secretion from virgin females strongly indicates a specific role of BC in combination with other components of the female ARG secretion.

The benefits of the use of anti-aphrodisiacs is obvious in *Pieris* species (Andersson et al. 2000; Andersson et al. 2003), as long as egg parasitoids are excluded. In *P. napi*, *P. rapae*, and *P. brassicae*, males synthesize volatiles and transfer them to the females during mating, which render the females unattractive to conspecific males. Thereby the substances reduce the rate of re-mating by the female. This unattractiveness benefits both the male and the female: males benefit by delaying female re-mating for as long as possible to avoid sperm competition from other males, whereas females benefit by not being harassed by other males and to have more time available for oviposition (Forsberg and Wiklund 1989). This time saving is confronted with the nutritional benefit of the spermatophores for the females. The value of the ejaculate gift is the higher the less often males have mated, thus to mate with virgin males benefits the females in this respect. However, our data suggest a disadvantage of mating with virgin males. They contain and probably transfer more anti-aphrodisiac to the females than already mated ones (N. E. Fatouros, unpublished data). This can lure egg parasitoids directly as a kairomone (Fatouros et al. 2005b) and indirectly by arresting them on the plant. Females might be able to "decide" whether they reject a courting male or not. If so, they could signal their receptivity to males that carry less anti-aphrodisiac to avoid eavesdropping egg parasitoids or prevent a plant's indirect defense response after egg deposition.

To our knowledge, this is the first report that males contribute to an elicitor of an oviposition-induced defensive plant response by transfer of BC to females during mating. BC is known to function intraspecifically as an anti-aphrodisiac that renders females less attractive to competing males. It is also exploited by the egg parasitoid *T. brassicae* directly as a kairomone (Fatouros et al. 2005b). Both roles of BC with respect to *Trichogramma* put the use of an anti-aphrodisiac by *P. brassicae* under selection pressure. As *Trichogramma* spp. are significant mortality factors for *P. brassicae* (van Heiningen et al. 1985), they might severely constrain the evolution of the butterflies' intraspecific sexual communication system.

Acknowledgements

We are grateful to Leo Koopman, Frans van Aggelen, and André Gidding for culturing the insects, and the experimental farm of Wageningen University (Unifarm) for breeding and providing the Brussels sprout plants. This work was supported by the Deutsche Forschungsgemeinschaft (Hi 416 / 15-1,2).

References

Alborn HT, Jones H, T., Stenhagen JG, Tumlinson JH (2000) Identification and synthesis of volaticin and related components from beet armyworm oral secretions. Journal of Chemical Ecology 26:203-220

Alborn HT, Turlings TCJ, Jones H, T., Stenhagen JG, Loughrin JH, Tumlinson JH (1997) An elicitor of plant volatiles from beet armyworm oral secretion. Science 276:945-949

Andersson J, Borg-Karlson A-K, Wiklund C (2000) Sexual cooperation and conflict in butterflies: a male-transferred anti-aphrodisiac reduces harassment of recently mated females. Proceedings of the Royal Society: Biological Sciences 267:1271-1275

Andersson J, Borg-Karlson A-K, Wiklund C (2003) Antiaphrodisiacs in Pierid butterflies: a theme with variation! Journal of Chemical Ecology 29:1489-1499

Bissoondath CJ, Wiklund C (1996) Male butterfly investment in successive ejaculates in relation to mating system. Behavioral Ecology and Sociobiology 39:285-292

Blaakmeer A, Hagenbeek D, van Beek TA, de Groot AE, Schoonhoven LM, van Loon JJA (1994a) Plant response to eggs vs. host marking pheromone as factors inhibiting oviposition by *Pieris brassicae*. Journal of Chemical Ecology 20:1657-1665

Blaakmeer A et al. (1994b) Isolation, identification, and synthesis of miriamides, new hostmarkers from eggs of *Pieris brassicae*. Journal of Natural Products 57:90-99

Boggs CL, Gilbert LE (1979) Male contribution to egg production in butterflies: Evidence for transfer of nutrients at mating. Science 206:83-84

Colazza S, Fucarino A, Peri E, Salerno G, Conti E, Bin F (2004a) Insect oviposition induces volatile emission in herbaceous plants that attracts egg parasitoids. Journal of Experimental Biology 207:47-53

Colazza S, McElfresh JS, Millar JG (2004b) Identification of volatile synomones, induced by *Nezara viridula* feeding and oviposition on bean spp., that attract the egg parasitoid *Trissolcus basalis*. Journal of Chemical Ecology 30:945-964

De Moraes CM, Meschner MC, Tumlinson JH (2001) Caterpillar-induced nocturnal plant volatiles repel conspecific females. Nature 410:577-580

Dicke M (1999) Are herbivore-induced plant volatiles reliable indicators of herbivore identity to foraging carnivorous arthropods? Entomologia Experimentalis et Applicata 91:131-142

Dicke M, van Loon JJA (2000) Multitrophic effects of herbivore-induced plant volatiles in an evolutionary context. Entomologia Experimentalis et Applicata 97:237-249

Fatouros NE, Bukovinszkine'Kiss G, Kalkers I A, Soler Gamborena R, Dicke M, Hilker M (2005a) Oviposition-induced plant cues: do they arrest *Trichogramma* wasps during host location? Entomologia Experimentalis et Applicata 115:207-215

Fatouros NE, Huigens ME, van Loon JJA, Dicke M, Hilker M (2005b) Chemical communication - Butterfly anti-aphrodisiac lures parasitic wasps. Nature 433:704

Forsberg J, Wiklund C (1989) Mating in the afternoon - time-saving in courtship and remating by females of a polyandrous butterfly *Pieris napi* L. Behavioral Ecology and Sociobiology 25:349-356

Gilbert LE (1976) Postmating female odor in *Heliconius* butterflies: A male-contributed antiaphrodisiac? Science 193:419-420

Gillott C (2002) Insect accessory reproductive glands: key players in production and protection of eggs. In: Hilker M, Meiners T (eds) Chemoecology of Insect Eggs and Egg Deposition. Blackwell Publishing LTd, Oxford, pp 37-59

Gillott C (2003) Male accessory gland secretions: Modulators of female reproductive physiology and behavior. Annual Review of Entomology:163-184

Halitschke R, Schittko U, Pohnert G, Boland W, Baldwin IT (2001) Molecular interactions between the specialist herbivore *Manduca sexta* (Lepidoptera, Spingidae) and its natural host *Nicotinia attenuata*. III. Fatty acid-amino acid conjugates in herbivore oral secretions are necessary and sufficient for herbivore-specific plant responses. Plant Physiology 125:711-717

Hilker M, Kobs C, Varama M, Schrank K (2002a) Insect egg deposition induces *Pinus sylvestris* to attract egg parasitoids. Journal of Experimental Biology 205:455-461

Hilker M, Meiners T (eds) (2002a) Chemoecology of Insect Eggs and Egg Deposition. Blackwell, Berlin

Hilker M, Meiners T (2002b) Induction of plant responses towards oviposition and feeding of herbivorous arthropods: a comparison. Entomologia Experimentalis et Applicata 104:181-192

Hilker M, Meiners T (2006) Early herbivore alert: insect egg induce plant defense. Journal of Chemical Ecology:in press

Hilker M, Rohfritsch O, Meiners T (2002b) The plant's response towards insect egg deposition. In: Hilker M, Meiners T (eds) Chemoecology of insect eggs and egg deposition. Blackwell Publishing, Berlin, Vienna, pp 205-233

Hilker M, Stein C, Schröder R, Varama M, Mumm R (2005) Insect egg deposition induces defence responses in *Pinus sylvestris*: characterisation of the elicitor. Journal of Experimental Biology 208:1849-1854

Klüsener B, Weiler E, W. (1999) Pore-forming properties of elicitors of plant defense reactions and cellulolytic enzyms. FEBS Letters 459:263-266

Leopold RA (1976) Role of Male Accessory Glands in Insect Reproduction. Annual Review of Entomology: 199-221

Mattiacci L, Dicke M, Posthumus MA (1995) β-Glucosidase: An elicitor of the herbivore-induced plant odor that attracts host-searching parasitic wasps. Proceedings of the National Academy of Sciences of the United States of America 92:2036-2040

Meiners T, Hilker M (1997) Host location in *Oomyzus gallerucae* (Hymenoptera: Eulophidae), an egg parasitoid of the elm leaf beetle *Xanthogaleruca luteola* (Coleoptera: Chrysomelidae). Oecologia 112: 87-93

Meiners T, Hilker M (2000) Induction of plant synomones by oviposition of a phytophagous insect. Journal of Chemical Ecology 26:221-232

Mumm R, Schrank K, Wegener R, Schulz S, Hilker M (2003) Chemical analysis of volatiles emitted by *Pinus sylvestris* after induction by insect oviposition. Journal of Chemical Ecology 29:1235-1252

Noldus LPJJ, van Lenteren JC (1985) Kairomones for the egg parasite *Trichogramma evanescens* Westwood II. Effect of contact chemicals produced by two of its hosts, *Pieris brassicae* L. and *Pieris rapae* L. Journal of Chemical Ecology 11:793-800

Nordlund DA, Strand MR, Lewis WJ, Vinson SB (1987) Role of kairomones from host accessory gland secretion in host recognition by *Telenomus remus* and *Trichogramma pretiosum*, with partial characterization. Entomologia Experimentalis et Applicata 44:37-44

Pohnert G, Jung V, Haukioja E, Lempa K, Boland W (1999) New fatty acid amides from regurgitant of lepidopteran (Noctuidae, Geometridae) caterpillars. Tetrahedron 55:11275-11280

Stowe MK, Turlings TCJ, Loughrin JH, Lewis WJ, Tumlinson JH (1995) The chemistry of eavesdroping, alarm and deceit. Proceedings of the National Academy of Sciences of the United States of America

92:23-28

Strand MR, Vinson SB (1982) Source and characterization of an egg recognition kairomone of *Telenomus heliothidis*; a parasitoid of *Heliothis virescens*. Physiological Entomology 7:83-90

Strand MR, Vinson SB (1983) Analyses of an egg recognition kairomone of *Telenomus heliothidis* (Hymenoptera: Scelionidae). Isolation and host function. Journal of Chemical Ecology 9:423-432

Truitt CL, Wei HX, Pare PW (2004) A plasma membrane protein from Zea mays binds with the herbivore elicitor volicitin. Plant Cell 16:523-532

Turlings TCJ, Alborn HT, Loughrin JH, Tumlinson JH (2000) Volicitin, an elicitor of maize volatiles in oral secretion of Spodoptera exigua: isolation and bioactivity. Journal of Chemical Ecology 26:189-202

van Heiningen TG, Pak GA, Hassan SA, van Lenteren JC (1985) Four year's results of experimental releases of *Trichogramma* egg parasites against lepidopteran pests in cabbage. Mededelingen van de Faculteit Landbouwwetenschappen Rijksuniversiteit Gent 50:379-388

Vet LEM, Dicke M (1992) Ecology of infochemical use by natural enemies in a tritrophic context. Annual Review of Entomology 37:141-172

Vinson SB (1984) How parasitoids locate their hosts: a case of insect espionage. In: Lewis T (ed) Insect Communication. Academic Press, London, pp 325-348

Wiklund C, Kaitala A, Lindfors V, Abenius J (1993) Polyandry and its effect on female reproduction in the green-veined white butterfly (*Pieris napi* L.). Behavioral Ecology and Sociobiology 33:25-33

7 The Response Specificity of *Trichogramma* Egg Parasitoids Towards Infochemicals During Host Location

N.E. Fatouros, G. Bukovinszkine'Kiss, M. Dicke & M. Hilker

Chapter 7

The Response Specifitcity of *Trichogramma* Egg Parasitoids Towards Infochemicals During Host Location

Abstract

Parasitoids are confronted with many different infochemicals of their hosts and food plants during host selection. Here, we investigated the effect of kairomones from the adult host *Pieris brassicae* and of cues present on Brussels sprout plants infested by *P. brassicae* eggs on the behavioral response of the egg parasitoid *Trichogramma evanescens*. Additionally, we tested whether the parasitoid's acceptance of *P. brassicae* eggs changes with different host ages. The wasps did not discriminate between olfactory cues from mated and virgin females or between mated females and males of *P. brassicae*. *T. evanescens* randomly climbed on the butterflies, showing a phoretic behavior without any preference for a certain sex. The parasitoid was arrested on leaf parts next to 1-day-old host egg masses. This arrestment might be due to cues deposited during oviposition. The wasps parasitized host eggs up to 3 days old equally well. Our results were compared with former studies on responses by the conspecific *T. brassicae* showing that *T. evanescens* makes less use of infochemicals from *P. brassicae* than *T. brassicae*.

Keywords: host finding, oviposition-induced plant synomone, phoresy, Brussels sprout plants, *Pieris brassicae*

Introduction

Chemical cues play a major role in the process of host selection by parasitoids (Vinson 1991), which has been divided into several steps such as habitat location, host location, and host acceptance. Variation in host location ability could be a major constraint for the performance of parasitoids used as pest-control agents (Lewis et al. 2003).

Egg parasitoids of herbivorous insects are known to use a wide range of chemicals for host selection, i.e. cues emitted by plants with and without host eggs, cues released by host adults, specifically the egg-laying female, and infochemicals of host eggs (Romeis et al. 2005; Rutledge 1996). *Trichogramma* egg parasitoids are considered efficient biological control agents and are used worldwide for control of lepidopteran pests in many crops (Smith 1996; Wajnberg and Hassan 1994). They are regarded as relatively polyphagous and therefore less host specific than specialist egg parasitoids such as some *Telenomus* spp. (Pinto and Stouthamer 1994). However, their suitability as biological control agents may vary due to considerable inter- and intraspecific variations in tolerance to environmental conditions, preference for hosts, recognition and acceptance of crops and host searching behavior (Pak 1988; van Dijken et al. 1986; Wajnberg and Hassan 1994). The ability to select and accept different host species may differ between *Trichogramma* strains and species (Pak et al. 1986; Pak and De Jong 1987; Pak et al. 1990). Evidence is accumulating that *Trichogramma* spp. show high habitat loyalty, thus resulting in development of preferences for specific hosts and specific plants in the respective habitat (see Romeis et al. 2005 and references therein).

Here, we investigated a tritrophic system consisting of Brussels sprout plants (*Brassica oleracea* var. *gemmifera* cv. Cyrus), the large cabbage white butterfly *Pieris brassicae* L. (Lepidoptera: Pieridae), and the egg parasitoid *Trichogramma evanescens* Westwood (Hymenoptera: Trichogrammatidae).

Egg deposition by *P. brassicae* has shown to induce changes in the plant surface chemistry of Brussels sprout plants after 3 days that result in the arrestment of females of *T. brassicae* Bezdenko (Fatouros et al. 2005a). The plant changes had arresting activity starting 3 days after egg deposition. Three-day-old eggs of *P. brassicae* were most successfully parasitized by *T. brassicae* (Fatouros et al. 2005a). While egg deposition by *P. brassicae* induced synomones perceived by contact, no evidence was found that *T. brassicae* is attracted to oviposition-induced plant volatiles (Fatouros et al. 2005a), as is known for several other egg parasitoids of herbivorous insects (Hilker and Meiners 2002; Hilker and Meiners 2006). However, volatiles from the adult hosts were shown to arrest *T. brassicae*. After locating mated *P. brassicae* females by the anti-aphrodisiac, benzyl cyanide, the wasps explicitly mount them and hitchhike with them to the host plant (i.e. phoresy) (Fatouros et al. 2005b).

The studied *T. evanescens* population was collected on *Mamestra brassicae* L. eggs (Lepidoptera: Noctuidae) found on *Brassica nigra*. Several strains of *T. evanescens* have been tested on their host acceptance for the three most harmful lepidopteran pest species in cabbage, i.e. *Mamestra brassicae, Pieris brassicae,* and *P. rapae* (van Dijken et al. 1986). The infochemicals known to be used by *T. evanescens* for host location of these cabbage pest species are: the host sex pheromone of *M. brassicae* and volatiles from virgin *P. brassicae* females (Noldus and van Lenteren 1985a), cues of their wing scales or other contact cues like chemicals from the host accessory gland (Noldus and van Lenteren 1985b). However, the identity of the used *Trichogramma* spp. was later revised to *T. maidis* (= *T. brassicae*) (Noldus 1989), which makes a comparison to these results difficult. Here, we used a population that was clearly identified as *T. evanescens* by internally transcribed spacer 2 (ITS2) DNA sequences (Stouthamer et al. 1999).

In the present study the response of the field-collected *T. evanescens* to specific host cues of the adult large cabbage white and to cues associated with an egg deposition on the plant, as well as their host acceptance of different *P. brassicae* egg ages was tested. These responses were compared to our earlier study on *T. brassicae* (Table 1). Since *T. evanescens* was found in *M. brassicae* eggs, we hypothesize that *T. evanescens* is less host specific in its response to infochemicals of *P. brassicae* than *T. brassicae* that was tested previously (Fatouros et al. 2005a).

Material and methods

Plants and herbivores. Brussels sprout plants were reared in a greenhouse (18 ± 2°C, 70% rh, L16:D8). Plants (8-12 weeks old) with ca. 14-16 leaves were used for rearing of *Pieris brassicae* and for the experiments.

Pieris brassicae was reared on Brussels sprout plants in a climate room (21 ± 1°C, 50-70% r.h., L16:D8). Each day a plant was placed into a large cage (80 x 100 x 80 cm) with more than 100 adults for approximately 24 hours to allow egg deposition.

Parasitoids. *Trichogramma evanescens* Westwood were collected from the field in *Mamestra brassicae* eggs in 2003 and reared since then in eggs of *P. brassicae* (25°C,

50-70% rh, L16:D8). For the rearing, 1 – 3-day-old *P. brassicae* eggs on leaves were used. Naïve wasps (no oviposition experience) were used for the host location tests with the adult hosts and for the host-age suitability tests. All two-choice contact bioassays with leaf disks were conducted with oviposition-experienced female wasps, because naïve *Trichogramma* wasps were shown to have a low response level toward plant cues (Fatouros et al. 2005a). An oviposition experience was given for a period of 18 h prior to the experiment with 2 - 3-day-old *P. brassicae* eggs deposited on Brussels sprout leaves. All wasps were mated and about 2 - 5 days old when tested. They were always provided with honey solution prior to the experiment.

Host suitability: oviposition tests with different host ages. *P. brassicae* eggs of 5 different ages (<12, 24, 48, 72 and 96 h) were offered on a leaf piece (ca. 1 mm^2) excised from egg-carrying plants (see procedure of egg infestation as described for the contact bioassays) to 1-day-old mated females of *T. evanescens*, which had had no previous contact to host eggs. An egg clutch consisting of 15 eggs of the same age was offered for a period of 24 h to a female confined in a small glass vial. After that period the wasp was removed from the vial. Eight females were tested per host age and parasitoid species. When the eggs turned black (approximately 4 days after exposure to the parasitoids), the number of parasitized eggs was counted. Each parasitized egg was isolated and individually kept in a gelatin capsule to determine the number of emerging wasps.

Host location: Effect of contact plant cues from egg-carrying plants. For the experiments, test plants were placed into the cage with more than 100 *P. brassicae* adults to allow deposition of eggs, wing scales, and host odors onto the plants for a period of 8 h. After this exposure to the butterflies the treated plants were either tested immediately or were kept in a climate chamber (21 ± 2°C, 70% rh, L16:D8) for 1 – 3 days longer. Thus, the period during which eggs or butterfly deposits were on the cabbage plant ranged from less than 12 h, to 24, 48, 72 or 96 h (compare Fatouros et al. 2005a). Control plants had never been in contact with *P. brassicae* or any other insect. They were grown under the same abiotic conditions as treated plants.

A test leaf square (2 cm^2) was cut from an egg-laden treated plant right next to an egg mass (3 - 5 cm away from the eggs). Such a test leaf square could elicit arrestment of wasps by either local oviposition-induced plant surface cues or by butterfly deposits such as scales that are always present close to eggs (Fatouros *et al.*, 2005a). A control square of the same size was cut from a leaf of corresponding position relative to the topmost leaves. A wasp was released in the center of a small glass Petri dish (5.5 cm diameter) lined with filter paper and simultaneously offered a test and a control leaf square. The total duration of time spent in the arena was observed for a period of 300 sec using The Observer software 3.0 (Noldus Information Technology 1993©). The time spent searching in the area outside the leaf squares was scored as "no response". Ten wasps were tested per day and plant. Test and control squares were changed after having tested 3 wasps. Each wasp was used only once and was then discarded.

Host location: Olfactometer bioassays with butterfly odors. The experiments were carried out in a two-chamber olfactometer described in detail by Fatouros et al. (2005a). This olfactometer was a modified version of the four-chamber olfactometer of Steidle and Schöller (1997). The time spent by the wasps in one of the two odor fields was observed for 300 sec. Two butterflies were introduced per chamber as odor source. A number of 10 - 15 naïve wasps was tested per day. To avoid biased results due to positional preferences of the parasitoids, the olfactometer was rotated 180° after every third insect. The response of *T. brassicae* to odor of the following *P. brassicae* combinations was tested: a) virgin females vs. mated females and b) mated females vs. males. A total number of 40 wasps per combination was tested.

Host location: Mounting bioassays. The selective mounting behavior of *T. evanescens* females was tested in a two-choice bioassay conducted at 23 ± 2 °C in a plastic container (9 cm high, 13.5 cm diameter). Two adults of *P. brassicae* were placed in the arena after cooling them down in a refrigerator (6°C) for ca. 10 min to decrease their mobility. A naïve *T. evanescens* female was introduced and then continuously observed till it made a first mount on one of the two butterflies. When a wasp did not choose for one of the two butterflies by mounting it within 5 min, this was recorded as "no response". The following butterfly combinations were tested: (a) mated females vs. males and (b) mated females vs. virgin females of *P. brassicae*. After each 10th wasp the butterflies were replaced by new ones. For each combination, 40 wasps were tested. Virgin female butterflies were obtained by separating pupae. Mated females had been observed to mate before they were used for the bioassay.

Statistics. Both the contact and the olfactometer bioassay were analyzed using Wilcoxon's matched pairs signed rank test. A χ^2-test was used to analyze the choices in the mounting bioassay. Parasitism rates for different host ages were analyzed with 5x2 contingency tables, and individual χ^2-tests were carried out, corrected by the sequential Bonferroni procedure for table-wide α levels.

Results

Host-age suitability. Wasps of *T. evanescens* did not show a preference for host eggs until the eggs were 72 hours old. About 50 % of the offered eggs were parasitized. However, eggs that were 96 h old were unsuitable for the wasps showing a low parasitization rate (Figure 1, 5x2 contingency test, df=4, P<0.001).

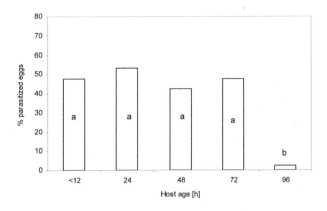

Figure 1: Host-age suitability of eggs from *Pieris brassicae* for *Trichogramma evanescens*. 15 eggs of different ages (<12, 24, 48, 72, 96 h) were offered on leaf pieces to 1-day old parasitoid females for 24 hours. Per egg age a number of 8 females was tested. Different letters in the columns indicate significant differences (P<0.05) (5x2 contingency tables using χ^2).

Effect of contact plant cues from egg-carrying plants. In order to examine whether plant surface chemicals in the vicinity of an egg mass serve as cues that indicate the close-by host eggs, leaf squares were excised right next to an egg mass and offered together with a leaf square from an egg-free "clean" leaf from a control plant. The leaf square from a plant, on which eggs had been deposited <12 h prior to the assay significantly arrested *T. evanescens* (Figure 2, P=0.03, Wilcoxon's matched pairs test). The same behavior was observed when leaf squares were offered from plants, on which eggs had been deposited 24 h prior to the assay. The wasps were significantly arrested on the test square (Figure 2, P=0.01, Wilcoxon's matched pairs test). Females of *T. evanescens* were not arrested on leaf squares cut from an egg-carrying leaf, on which eggs had been deposited 48 h or longer prior to the assay. In all experiments, the wasps showed a relatively low response and spent about 50 % or more of the total time in the "no response" area in all tested treatments (Figure 2).

Response to butterfly odors. In two-choice olfactory bioassays the *T. evanescens* wasps showed no preferences for odors of *P. brassicae* females or males (Figure 3). When odor from virgin female butterflies was offered against odor of mated female butterflies, the wasps did not show any preference either.

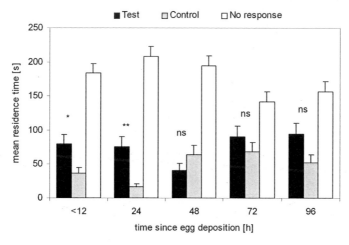

Figure 2: Contact 2-choice bioassay. Response of *Trichogramma evanescens* females to test leaf parts right excised from a leaf next to an egg mass deposited <12, 24, 48, 72, 96 h prior to the bioassay (black bars) and to egg-free control leaf parts (grey bars). White bars: Time spent by wasps on bioassay surface other than leaves (i.e. "no response"). Number of tested females per treatment N=50. Mean residence time and standard error are shown. Asterisks indicate significant differences between test and control within the same treatment *P<0.05, **P<0.01, ns, not significant (Wilcoxon's matched pairs signed rank test).

Mounting of butterflies. We exposed adult butterflies to female wasps in two-choice bioassays. Indeed, the wasps mounted adult butterflies. However, the *T. evanescens* did not discriminate between mated females, virgin female or male butterflies (Figure 3).

Discussion

Trichogramma evanescens did not show any preference for volatile cues emitted by adult *P. brassicae*. They showed phoretic behavior by climbing onto the butterflies but did not discriminate between males and females or mated and virgin females. Since such a mounting behavior involves some risk from the butterfly defending itself, it is unlikely that this behavior just coincidently happened. Leaf disks of egg-laden plants arrested the wasps until 24 h after egg deposition. Here, a leaf contamination with host cues like wing scales, always deposited in the vicinity of eggs, could have caused this increased searching behavior. An arrestment to such

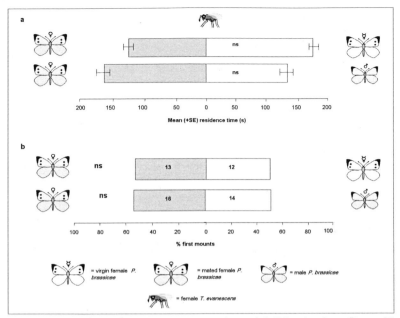

Figure 3: Response of *T. evanescens* wasps to cues from adult *P. brassicae* butterflies. **a.** Mean residence time (± SE) spent by wasps in test and control fields of a two-chamber olfactometer; n=40 tested wasps per experiment; ns, not significant (Wilcoxon's matched pairs signed ranks test). **b.** Proportion of first mounts (%) of wasps on adult *P. brassicae* butterflies. The number of responding wasps is shown inside each bar; n=40 wasps tested per combination; ns, not significant (χ^2-Test).

deposits on a host plant was shown for *T. brassicae* (Fatouros et al. 2005a). Jones et al. (1973) showed that *T. evanescens* is able to detect tricosane from adult host scales (*Heliothis zea*), which stimulates parasitization. In our study, leaf disks from plants with eggs and deposits older than 24 h did not arrest the wasps. Host eggs that had been offered to the wasps were equally parasitized until the age of 72 h.

There are interesting parallels and differences between the responses of *T. evanescens* and *T. brassicae* to egg-laden leaves and host cues (compare Table 1):

(a) Like *T. evanescens*, *T. brassicae* is arrested on leaf areas right next to *P. brassicae* eggs. The arrestment response vanishes 24 h after egg deposition in both *T. brassicae* and *T. evanescens* (Fatouros et al. 2005a).

(b) However, 72 h after egg deposition, *T. brassicae* was again arrested on leaf area next to an egg mass. A series of different experiments indicated that the eggs of *P. brassicae* induce a change in the leaf surface of Brussels sprout plants arresting *T. brassicae* wasps at a time when eggs are most suitable for parasitization, i.e. 72 h after egg deposition (Fatouros et al. 2005a). Such a response to oviposition-induced plant surface changes has not been observed in *T. evanescens*.

(c) *T. evanescens* was not arrested by cues from mated females. In contrast, *T. brassicae* preferred mated over virgin females, which was mediated by benzyl cyanide, the anti-aphrodisiac of *P. brassicae* that is transferred from males to females during mating to render the females less attractive to conspecific males (Andersson et al. 2003).

Table 1: Response of females of *T. brassicae* and *T. evanescens* to infochemicals of their host *P. brassicae* and Brussels sprout plants infested with *P. brassicae* eggs.

Infochemical	Source		Response	
			T. brassicae	*T. evanescens*
Volatile	Plant	Egg-laden (24 - 72 h)	No response [a]	*Not tested*
	Host	Virgin female	No response [b]	No discrimination
		Mated female	Arrestment [b]	No discrimination
		Males	Arrestment [b]	No discrimination
Contact/	Plant	Egg-laden (up to 24 h)	Arrestment [a]	Arrestment
close range		Egg-laden (72 - 96 h)	Arrestment [a]	No response
	Host	Virgin female	Not attractive [b]	No discrimination
		Mated female	Mounting [b]	No discrimination
		Males	Not attractive [b]	No discrimination
		Scales/deposits (24-h-old)	Arrestment [a]	*Not tested*
		Scales/deposits (72-h-old)	No response [a]	*Not tested*

[a] Fatouros et al. 2005a
[b] Fatouros et al. 2005b

Even though both *T. evanescens* and *T. brassicae* are known as host generalists, at least from laboratory studies, that readily accept *P. brassicae* as host, we found such species specific differences in their responsiveness to host cues relevant for host location. How to explain these differences?

T. evanescens has been collected from *Mamestra brassicae* eggs on *Brassica nigra* and was reared for a shorter period on *P. brassicae* than *T. brassicae*. The latter was collected from *E. kuehniella* eggs placed on cardboard cards in a cabbage garden. Even though the original host of *T. brassicae* is unknown, the longer rearing period on *P. brassicae* might have resulted in a stronger adaptation to cues of *P. brassicae* or plant cues induced by this species. The longer rearing period might have caused a stronger selection of those *T. brassicae* individuals that are most sensitive to *P. brassicae* cues. Kaiser et al. (1989) showed that the affinity for an initially non-preferred host was enhanced when *Trichogramma* was reared on this species. Such an enhancement might also be due to pre-imaginal conditioning (Bjorksten and Hoffmann 1995; Kaiser et al. 1989) or adult-experience (Bjorksten and Hoffmann 1998; Kaiser et al. 1989; van Dijken et al. 1986), if rearing periods are too short to allow selective effects. However, this is not the case when comparing *T. evanescens* and *T. brassicae*. The latter species was reared at least for 30 generations longer on *P. brassicae*. When studying the selective power of rearing conditions, already Pimentel *et al.* (1967) could show that oviposition preferences may shift more and more to a new host with an increasing number of generations reared on this new host.

However, host preferences of generalist parasitoids may also be stable at varying rearing conditions, indicating very conservative and robust sensitivities to host cues. *T. maidis* (= *T. brassicae*) had a greater spontaneous affinity for *Ostrinia nubilalis* than for *Ephestia kuehniella* with respect to their host acceptance (Kaiser et al. 1989). This preference was based on conservative genetic determination, because it survived the rearing of the parasitoid for more than 100 generations on the non-preferred host. Similar results were observed for other *Trichogramma* spp. as well (Pak 1988). Chemical parameters, like surface kairomones, play a role in the specific recognition of host eggs. Thus, *Trichogramma* spp. may have an innate ability to use specific kairomones of certain hosts during the host selection procedure.

There is growing evidence for *Trichogramma* spp. "being more prevalent in certain habitats or on specific plants" (Romeis et al. 2005). Such a habitat / plant loyalty might be due to their limited moving abilities. Like the effects of laboratory rearing, habitat / plant loyalty might change the wasps´ responsiveness to certain host and plant cues. Egg parasitoids living in monocultures or habitats of little biodiversity would encounter only a small range of host species when they do not leave these sites. Therefore, some *Trichogramma* spp. with high habitat / plant loyalty might show a higher specificity with respect to the infochemicals necessary for host location than generally assumed. They could still keep the ability to parasitize most host egg species as soon as encountered, so that they remain generalists with respect to the host acceptance.

The response of closely related parasitoid species to infochemicals of the same host has been compared only in a few studies. Bruni et al. (2000) tested the kairomonal activity of the attractant pheromone of the spined soldier bug, *Podisus maculiventris* (Hemiptera: Pentatomidae) on two *Telenomus* species, one a generalist egg parasitoid of pentatomids (*T. podisi*) and the other a phoretic specialist on *Podisus* eggs (*T. calvus*). There was no evidence that *T. podisi* uses the pheromone as a kairomone. In contrast, *T. calvus* females oriented to the volatile chemicals of their host, to find areas likely to contain hosts and to use the pheromone to guide their phoretic behavior. Here, the dietary breadths of the two *Telenomus* wasps are different and were likely to explain their different response toward the specific host cue.

In conclusion, our results do not provide evidence that *T. evanescens* responds to volatile infochemicals released from adult *P. brassicae* or from leaves with eggs of *P. brassicae*. However, *T. evanescens* does accept *P. brassicae* eggs for parasitization. This non-response to volatile infochemicals of *P. brassicae* and its host plant Brussels sprouts contrasts to the response of the closely related *T. brassicae*, which was shown to innately use the anti-aphrodisiac benzyl cyanide of mated *P. brassicae* females and to learn oviposition-induced plant cues after landing on the host plant (Fatouros et al. 2005a; b). Thus, the two polyphagous egg parasitoids with very similar host ranges respond very differently to host and plant infochemicals. Further studies are needed to elucidate whether these differences are specific for a *Trichogramma* species or whether such differences can be detected even between *Trichogramma* populations with high, but different habitat / plant loyalty.

Acknowledgements

The authors thank Katja Hoedjes and Linda Tiggelman for their help with some of the experiments, Martinus E. Huigens for comments on a previous version of the manuscript, Leo Koopman, Frans van Aggelen, and André Gidding for culturing the insects, and the experimental farm of Wageningen University (Unifarm) for breeding and providing the Brussels sprout plants. The study was financially supported by the Deutsche Forschungsgemeinschaft (Hi 416/15-1, 2).

References

Andersson J, Borg-Karlson A-K, Wiklund C (2003) Antiaphrodisiacs in Pierid butterflies: a theme with variation! Journal of Chemical Ecology 29:1489-1499

Bjorksten TA, Hoffmann AA (1995) Effects of pre-adult and adult experience on host acceptance in choice and non-choice tests in two strains of *Trichogramma*. Entomologia Experimentalis et Applicata 76:49-58

Bjorksten TA, Hoffmann AA (1998) Persistence of experience effects in the parasitoid *Trichogramma* nr. *brassicae*. Ecological Entomology 23:110-117

Bruni R, Sant'Ana J, Aldrich JR, Bin F (2000) Influence of host pheromone on egg parasitism by scelionid wasps: Comparison of phoretic and nonphoretic parasitoids. Journal of Insect Behavior 12:165-173

Fatouros NE, Bukovinszkine'Kiss G, Kalkers LA, Soler Gamborena R, Dicke M, Hilker M (2005a) Oviposition-induced plant cues: do they arrest *Trichogramma* wasps during host location? Entomologia Experimentalis et Applicata 115:207-215

Fatouros NE, Huigens ME, van Loon JJA, Dicke M, Hilker M (2005b) Chemical communication - Butterfly anti-aphrodisiac lures parasitic wasps. Nature 433:704

Hilker M, Meiners T (2002) Induction of plant responses towards oviposition and feeding of herbivorous arthropods: a comparison. Entomologia Experimentalis et Applicata 104:181-192

Hilker M, Meiners T (2006) Early herbivore alert: insect eggs induce plant defense. Journal of Chemical Ecology:in press

Jones RL, Lewis WJ, Beroza M, Bierl BA, Sparks AN (1973) Host-seeking stimulants (kairomones) for the egg parasite, *Trichogramma evanescens*. Environmental Entomology 2:593-596

Kaiser L, Pham-Delegue MH, Masson C (1989) Behavioural study of plasticity in host preferences of *Trichogramma maidis* (Hym.: Trichogrammatidae). Physiological Entomology 14:53-60

Lewis WJ, Vet LEM, Tumlinson JH, van Lenteren JC, Papaj DR (2003) Variations in natural-enemy foraging behaviour: Essential element of a sound biological control theory. In: Van Lenteren JC (ed) Quality Control and Production of Biological Control Agents: Theory and Testing Procedures. CAB Publisher, Wallingford, pp 41-58

Noldus LPJJ (1989) Semiochemicals, foraging behaviour and quality of entomophagous insects for biological control. Journal of Applied Entomology 108:425-451

Noldus LPJJ, van Lenteren JC (1985a) Kairomones for the egg parasite *Trichogramma evanescens* Westwood I. Effect of volatile substances released by two of its hosts, *Pieris brassicae* L. and *Mamestra brassicae*, L. Journal of Chemical Ecology 11:781-791

Noldus LPJJ, van Lenteren JC (1985b) Kairomones for the egg parasite *Trichogramma evanescens* Westwood II. Effect of contact chemicals produced by two of its hosts, *Pieris brassicae* L. and *Pieris rapae* L. Journal of Chemical Ecology 11:793-800

Pak GA (1988) Selection of *Trichogramma* for inundative biological control. In. Wageningen Agricultural University, Wageningen, The Netherlands, p 224

Pak GA, Buis H, Heck ICC, Hermans MLG (1986) Behavioral Variations Among Strains Of *Trichogramma* Spp - Host-Age Selection. Entomologia Experimentalis Et Applicata 40:247-258

Pak GA, De Jong EJ (1987) Behavioral variations among strains of *Trichogramma* spp - Host recognition. Netherlands Journal Of Zoology 37:137-166

Pak GA, Kaskens JWM, Dejong EJ (1990) Behavioral variation among strains of *Trichogramma* spp - Host-species selection. Entomologia Experimentalis Et Applicata 56:91-102

Pimentel D, Smith GJC, Soans J (1967) A population model to sympatric speciation. American Naturalist 101:493-504

Pinto JD, Stouthamer R (1994) Systematics of Trichogrammatidae with emphasis on *Trichogramma*. In: Wajnberg E, Hassan SA (eds) Biological Control with Egg parasitoids. CAB international, Oxon, pp 1-36

Romeis J, Babendreier D, Wäckers FL, Shanower TG (2005) Habitat and plant specificity of *Trichogramma* egg parasitoids - underlying mechanisms and implications. Basic and Applied Ecology 6:215-236

Rutledge CE (1996) A survey of identified kairomones and synomones used by insect parasitoids to locate and accept their hosts. Chemoecology 7:121-131

Smith SM (1996) Biological control with *Trichogramma*: Advances, successes, and potential of their use. Annual Review of Entomology 41:375-406

Steidle JLM, Schoeller M (1997) Olfactory host location and learning in the granary weevil parasitoid *Lariophagus distinguendus* (Hymenoptera: Pteromalidae). Journal of Insect Behavior 10:331-342

Stouthamer R, Hu J, Kan vFJPM, Platner GR, Pinto JD (1999) The utility of internally transcribed spacer 2 DNA sequences of nuclear ribosomal gene for distinguishing sibling species of *Trichogramma*. BioControl 43:421-440

van Dijken MJ, Kole M, van Lenteren JC, Brand AM (1986) Host-preference studies with *Trichogramma evanescens* Westwood (Hym., Trichogrammaticdae) for *Mamestra brassicae*, *Pieris brassicae* and *Pieris rapae*. Journal Applied Entomology 101:64-85

Vinson SB (1991) Chemical signals used by parasitoids. In: Bin F (ed), vol. 74, 3 edn, Perugia, pp 15-42

Wajnberg E, Hassan SA (1994) Biological Control with Egg Parasitoids. CAB International, Wallingford, Oxon, UK

8 Herbivory-Induced Plant Volatiles Mediate In-flight Host Discrimination by Parasitoids

N.E. Fatouros, J.J.A. van Loon, K.A. Hordijk, H.M. Smid & M. Dicke

Chapter 8

Herbivory-Induced Plant Volatiles Mediate In-Flight Host Discrimination by Parasitoids

Abstract

Herbivore feeding induces plants to emit volatiles that are well-detectable and reliable cues for foraging parasitoids which allow them to perform oriented host searching. We investigated whether these plant volatiles play a role in avoiding parasitoid competition by discriminating parasitized from unparasitized hosts in flight. In a wind tunnel set-up we used mechanically damaged plants treated with regurgitant containing elicitors to simulate and standardize herbivore feeding. The solitary parasitoid *Cotesia rubecula* discriminated between volatile blends from Brussels sprouts plants treated with regurgitant of unparasitized *Pieris rapae* or *P. brassicae* caterpillars over blends emitted by plants treated with regurgitant of parasitized caterpillars. The gregarious *Cotesia glomerata* discriminated between volatiles induced by regurgitant from parasitized and unparasitized caterpillars of its major host species, *P. brassicae*. GC-MS analysis of headspace odors revealed that cabbage plants treated with regurgitant of parasitized *P. brassicae* caterpillars emitted lower amounts of volatiles than plants treated with unparasitized caterpillars. We demonstrate (1) that parasitoids can detect, in flight, whether their hosts contain competitors, and (2) that plants reduce the production of specific herbivore-induced volatiles after a successful recruitment of their bodyguards. As the induced volatiles bear biosynthetic and ecological costs to plants, down-regulation of their production has adaptive value. These findings add a new level of intricacy to plant-parasitoid interactions.

Keywords: induced plant defense, headspace analysis, tritrophic interactions, *Pieris*, *Cotesia*

Introduction

Female parasitoids of herbivorous insects have to search for hosts to lay their eggs in or on them. Host searching is known to be guided by volatile synomones that are mostly produced by the host plants after induction caused by herbivore feeding (Dicke 1999; Hilker and Meiners 2002; Shiojiri et al. 2001; Steidle and van Loon 2003; Turlings et al. 2002). The ability to distinguish unparasitized from parasitized hosts and to reject the latter for oviposition (i.e. host discrimination), is widespread among hymenopteran parasitoids (van Alphen and Visser 1990; van Lenteren 1981; Vinson 1985). The reproductive success of endoparasitoids is dependent on the quality and size of the hosts, because hosts provide a limited nutritional resource for their offspring (Godfray 1994). Oviposition on or into already parasitized hosts by a conspecific parasitoid (i.e. superparasitism) can lead to competition with deleterious consequences such as mortality or reduced fitness. Therefore, avoiding superparasitism can have several advantages, including the optimization of foraging efficiency (Godfray 1994; van Lenteren 1981).

So far, host discrimination by parasitoids is known to be mediated by marking pheromones deposited by the ovipositing female either on or in the host, or within the explored patch. In addition, physiological changes in the host hemolymph induced by the parasitoid progeny (reviewed by Nufio and Papaj 2001; van Lenteren 1981) or physical changes of the host surface acting as external markers (Takasu and Hirose 1988) may provide cues for host discrimination. Marking pheromones are usually perceived through contact chemoreception (van Lenteren 1981) although some examples of olfactory perception are known as well (Sheehan et al. 1993; van Baaren and Nenon 1996; van Giessen et al. 1993).

Herbivore-induced volatiles are emitted in considerable quantities by plants, and the blend induced by herbivory differs in a qualitative and/or quantitative sense from the volatile blends of undamaged or mechanically damaged plants (Dicke and Vet 1999; Shiojiri et al. 2000; Turlings et al. 1995). The emitted blend can be specific for the plant species, the herbivore species that damages the plant, and even for a specific stage of the herbivore (reviewed by Dicke and Vet 1999; Takabayashi and Dicke 1996; Turlings et al. 1995). Based on the high detectability and reliability of herbivore-induced plant volatiles, parasitoids are able to use these cues to locate their hosts from a distance (Vet and Dicke 1992). However, until now it is unknown whether foraging females are able to distinguish unparasitized from parasitized hosts at a distance via olfactory cues produced by their host plant. Yet, this ability would save them foraging time and would reduce risks associated with host defense. We have investigated this for a tritrophic system consisting of *Brassica* plants, *Pieris* caterpillars, and *Cotesia* parasitoids.

Brussels sprouts plants are known to emit a blend of volatiles when infested by *Pieris brassicae* L. (Lepidoptera: Pieridae) or *P. rapae* L. caterpillars. These induced volatiles attract *Cotesia rubecula* Marshall (Hymenoptera: Braconidae), a solitary endoparasitoid of the small cabbage white *Pieris rapae* and *C. glomerata* L., a gregarious parasitoid of both *Pieris* species (Agelopoulos and Keller 1994; Geervliet et al. 1994; Geervliet et al. 1998; Mattiacci et al. 1995; Mattiacci et al. 2001). Differences between headspace volatiles collected from undamaged and *Pieris*-infested Brussels sprouts plants have been extensively documented (Blaakmeer et al. 1994; Mattiacci et al. 1995; Mattiacci et al. 2001; Smid et al. 2002). Coupled gas chromatography-electroantennography (GC-EAG) analysis revealed that 20 of the headspace volatiles elicit a response in the antennae of *C. rubecula* and *C. glomerata* (Smid et al. 2002). The application of regurgitant of *Pieris brassicae* larvae or the enzyme β-glucosidase into a mechanical wound elicits induction of similar volatile blends in cabbage plants as herbivory, showing that the plants recognize infestation through an elicitor in the oral secretion of *Pieris* caterpillars (Mattiacci et al. 1995).

In the present study we investigate (1) whether females of the two *Cotesia* species prefer, in flight, induced odors emitted by plants fed upon by unparasitized caterpillars or treated with their regurgitant over odors emitted by plants fed upon by parasitized caterpillars or treated with their regurgitant, and (2) whether the induced plant odor blends differ, thereby providing a chemical basis for this discrimination behavior in the two wasp species.

Material and methods

Plants and insects. Brussels sprouts plants (*Brassica oleracea var. gemmifera* cv. Cyrus) were reared in a greenhouse (18 ± 2 °C, 70% RH, 16L: 8D). Plants (8 – 12-wk old) having ca. 14 - 16 leaves were used for the experiments.

Pieris rapae and *P. brassicae* were reared on Brussels sprouts plants in a climate room (21 ± 1 °C, 50-70% RH, 16L: 8D). The parasitoid wasps *Cotesia rubecula* and *C. glomerata* were reared on *P. rapae* and *P. brassicae* larvae, respectively, feeding on Brussels sprouts plants under greenhouse conditions (see above).

For bioassay experiments *C. rubecula* and *C. glomerata* pupae were collected and reared in gauze cages in a climate room (23 ± 1 °C, 50-70% RH, 16L: 8D). Once eclosed, the wasps were provided with water and honey. They had no contact with plant material or caterpillars before the initiation of the bioassays. They are referred to as naive wasps.

Regurgitant collection. Regurgitant was collected from unparasitized and parasitized 5th instar larvae of *P. brassicae*, and from unparasitized and parasitized 4th instar *P. rapae* larvae 24 hr before the bioassay or headspace collection. The regurgitant droplets were collected with a 5 µl glass capillary tube after gently squeezing the caterpillars with forceps. About 5 - 10 caterpillars of each group (parasitized and unparasitized larvae) were used to collect 50 µl regurgitant, which was immediately put in separate vials on ice. Larvae that had been exposed to oviposition by female wasps were dissected before the regurgitant was pooled to ensure that they were indeed parasitized.

Plant treatments. For the bioassay with *P. rapae* caterpillars, 30 1st instar caterpillars were offered to females of *C. rubecula*. Each caterpillar was observed to be stung by a wasp and subsequently reared on cabbage. Another 30 unparasitized L1 caterpillars were reared on cabbage under identical circumstances. After 4 days, 15 unparasitized larvae were placed on a Brussels sprouts plant. Another 15 larvae that were supposed to be parasitized were placed on a different plant. After 48 hr the plants were used in the bioassay, to investigate parasitoid odor preference. Parasitized caterpillars were reared until the last instar, to examine whether they had indeed been parasitized and to assess the degree of parasitization.

In bioassays during which caterpillar regurgitant was used, each plant was damaged on the 5th fully expanded leaf and for headspace collection on the 5th and 6th leaf from the top by using a pattern wheel. The wheel was rolled over the leaf surface on each side of the mid rib, 10 lines in parallel, and 10 lines perpendicularly to the mid rib creating a ca. 8 cm² field with punctures being distributed over about 1/3 of the leaf surface. Per artificially damaged leaf 25 µl of regurgitant, freshly collected from either parasitized or unparasitized caterpillars, was applied with a micropipette and distributed with a brush. Afterwards the entire plants were kept in a climate room (23 ± 1 °C, 50-70% RH) for 18 to 24 hr until the bioassay or headspace analysis. After the induction period only the treated leaves were used for both behavioral bioassays and headspace analysis.

Bioassays. Parasitoid odor preference experiments were conducted in a windtunnel set-up (25±5°C, 50-70% r.h., 0.7 kLux) described by Geervliet et al. (1994), with a wind speed of 0.2 m/s. For the bioassays only the treated leaves, that were of the same physiological age and size, were used and excised from the induced plants just prior to the experiment. In this way a highly standardized set of odor sources was offered to the parasitoids in a choice situation (Mattiacci et al. 1994; Mattiacci et al. 1995). The petioles were inserted in vials with water and the two leaves acting as odor sources were placed at the upwind end of the windtunnel, ca. 14 cm apart. Experiments were carried out on 9 - 12 d, spread out over 2 - 3 months for each treatment. For each of these days new plants and new wasps were used. Three different plant treatments were tested in a two-choice flight experiment in the windtunnel: (1) cabbage leaves infested with either parasitized or unparasitized 2nd instar larvae of *P. rapae*, (2) artificially damaged cabbage leaves treated with regurgitant collected from either unparasitized (UNPAR) or parasitized (PAR) *P. rapae*, and (3) artificially damaged cabbage leaves treated with regurgitant from either unparasitized or parasitized *P. brassicae* caterpillars. Two to 6-d-old, naive female wasps of each species were used. Each wasp was individually released on a microscope slide with a small part of a leaf from a caterpillar-damaged cabbage plant from which first instar caterpillars had been removed just prior to the bioassay. For each *Cotesia* species damage done by the *Pieris* species under test was used. This priming served to increase the responsiveness of the wasps during the bioassay. The microscope slide was placed in the middle of the release cylinder, which was 60 cm downwind from the odor sources.

After release the behavior of the wasp was observed. Only flights that resulted in the first landing on one of the odor sources were recorded as response. All landings on other parts of the windtunnel besides the release cylinder or odor source were recorded as no response. If the wasp remained longer than 10 min in or on the release cylinder it was recorded as no response and discarded thereafter. During the bioassay, plant positions were changed from left to right and vice versa after every second wasp. Females of the two wasp species were alternately tested for every treatment with a maximum number of 15 females per species and day. The two species were tested on the same day to expose them to similar conditions. On each observation day one leaf from one plant of each treatment was tested.

Collection of headspace volatiles. Headspace collection was performed at 23±1°C, 50-70 % r.h,. and approximately 10 klux; 9 replicate plants were used per treatment. Volatiles from the treatments to be compared were collected on the same day to minimize variation among plant and caterpillar batches as well as day-to-day variation. Four replicates were obtained from the same batch of 10 – 12-wk-old plants within 9 d in September 2002, and five replicates were obtained from the same batch of ca. 10 – 12-wk-old plants within 11 d in April 2003. First, incoming air was led through a 5 l glass jar for 1 hr, at a flow rate of 300 ml/min to clean the system.

Then, two leaves of the same plant induced by regurgitant from unparasitized *P. brassicae* caterpillars, and two leaves from a different plant induced by regurgitant of parasitized caterpillars were excised and immediately placed into separate jars with their petioles in a vial with water. They were purged for a further 1 hr. During the 3rd hr the headspace was collected using traps (10 cm long), packed with 200 mg Tenax TA (Markes, Pontyclun, UK, 60-80 Mesh) connected to the air outlet.

GC-MS analysis of headspace samples. Volatiles were desorbed from the Tenax traps using an automated thermodesorption unit (model Unity, Markes, Pontyclum, United Kingdom) at 200 °C for 10 min (He flow 30 ml/min) and focused in an electrically cooled (−10 °C) sorbent trap to allow for narrow starting bands. Volatiles were injected into the GC in splitless mode by ballistic heating of the cold trap for 3 min to 270 °C. After separation by capillary gas chromatography (column: 30 m x 0.32 mm ID RTX-5 Silms, film thickness 0.33 μm; temperature program: 40 °C to 95 °C at 3°C/min, then to 165 °C at 2 °C/min, and finally to 250 °C at 15 °C/min) volatiles were introduced into a Finnigan quadrupole mass spectrometer (Model Trace) operating at 70 eV in EI ionisation mode. Mass spectra were recorded with full scan mode (33-300 AMU, 3 scans per second). Compounds were identified using deconvolution software (AMDIS) in combination with Nist 98 and Wiley 7th edition spectral libraries and by comparing their linear retention indices with those of authentic references from our own library.

Other identifications were made by comparison of mass spectra and linear retention indices (Adams 2001; Ruther 2000), by interpolating homologous series and/or by using reference substances. The identified peaks were integrated by Xcalibur software (Finnigan). To minimize any interference by co-eluting compounds, specific ions were selected for each compound of interest. Generally, these ions were identical with the model ions indicated by AMDIS.

Statistics. The null hypothesis (H_0) was that there is no difference in the probability of wasps flying to the UNPAR treatment or PAR treatment. The alternative hypothesis (H_1) was that the wasps used herbivore-induced plant volatiles in host-discrimination, and thus that they preferred volatiles from the UNPAR treatment over those from the PAR treatment. The choices of the wasps in the 2-choice olfactometer conform to a binomial distribution. Therefore, choices of the wasps between two odor sources within a bioassay were analyzed using a one-sided binomial test (Siegel 1956). Responsiveness among the choice tests between the different treatments was analyzed with 2 x 2 contingency tables using chi-square-statistics.

Amounts of volatiles trapped were analyzed on the basis of peak area, as determined in the GC-MS analysis. For the headspace analyses, samples of both treatments collected on the same day were analyzed as paired data. The Wilcoxon's matched pair signed rank test was employed to test whether the peak area per compound differed between leaves treated with regurgitant collected from parasitized and unparasitized caterpillars. A sign test was used to determine whether the number of compounds that were emitted in larger amounts differed from a 50:50 distribution over the two treatments.

Results

Parasitoid behavior. Females of *C. rubecula* landed significantly more often on leaves infested with unparasitized larvae of *P. rapae* than on leaves infested with parasitized caterpillars (Fig. 1a; $P = 0.01$). Only those combinations of leaves were taken into account for which the parasitisation rate after dissection was found to be minimally 50 %.

Because the amount of leaf surface area removed from leaves fed upon by the unparasitized caterpillars was about 20 % larger than from leaves fed upon by parasitized caterpillars, and amount of *ad libitum* feeding could not be manipulated, we conducted subsequent bioassays with leaves to which a standardized amount of artificial damage and regurgitant were applied. In bioassays with regurgitant of *P. rapae* larvae, females of *C. rubecula* significantly preferred to land on leaves treated with regurgitant of unparasitized caterpillars (Fig. 1b; $P = 0.039$). Also when leaves treated with regurgitant of *P. brassicae* were offered, *C. rubecula* females landed significantly more often on the UNPAR treatment than on the PAR treatment (Fig. 1c; $P = 0.045$).

C. glomerata wasps did not discriminate between artificially damaged cabbage leaves treated with regurgitant collected from either unparasitized or parasitized *P. rapae* larvae (Figure 2a). However, in tests using leaves treated with regurgitant of *P. brassicae* larvae, *Cotesia glomerata* females showed a clear preference for landing on leaves treated with regurgitant of unparasitized larvae (Fig. 2b; $P = 0.002$). Additionally, the wasps´responsiveness was significantly higher in the tests with leaves treated *P. brassicae* regurgitant than in tests with regurgitant of *P. rapae* larvae ($P < 0.001$, $\chi^2 = 17.2$; df =1).

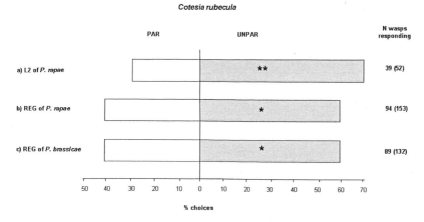

Figure 1: Response of *C. rubecula* to cabbage leaves a) exposed to parasitized (PAR) and unparasitized (UNPAR) 2nd instar larvae of *P. rapae*, b) artificially damaged and treated with regurgitant (REG) of PAR and UNPAR 4th instar *P. rapae* larvae, and c) artificially damaged and treated with REG from PAR and UNPAR 5th instar *P. brassicae* larvae in a two-choice set-up. Asterisks indicate significant differences within a choice test: * $P < 0.05$, ** $P < 0.01$ (binomial test). The bars show the percentages of wasps landing on either odor source. The numbers show the number of wasps responding; the numbers in brackets show the total number of wasps tested. All wasps were released on a leaf damaged by first instar larvae either of UNPAR *P. rapae* (a/b) or UNPAR *P. brassicae* (c).

Headspace analysis. We focused our headspace analysis on those compounds that elicited electrophysiological responses from antennal olfactory receptors of the two *Cotesia* species (Smid et al. 2002). The plants treated with regurgitant of parasitized or unparasitized *P. brassicae* larvae emitted a volatile blend that consisted of green leaf volatiles, aldehydes, alcohols, and esters, terpenes and methyl salicylate. Averaged over 9 replications, in total 19 out of the 22 compounds were emitted in larger amounts in the UNPAR treatment than in the PAR treatment ($P < 0.001$, sign test). In addition, significantly higher amounts of the major terpenoids limonene ($P = 0.05$), α-thujene ($P = 0.04$), and of 1,8-cineole ($P = 0.01$; two-sided Wilcoxon's matched pair signed rank test) were released in the UNPAR treatment (Figure 3). Thus, the quantitative profile of the emitted volatiles changes and three volatiles are emitted in significantly lower amounts when cabbage plants are induced by regurgitant of parasitized *P. brassicae* compared to cabbage plants induced by regurgitant of unparasitized *P. brassicae*.

Discussion

Our results show that *Cotesia rubecula* females without oviposition experience discriminated suitable hosts, i.e. hosts that were not parasitized, from parasitized

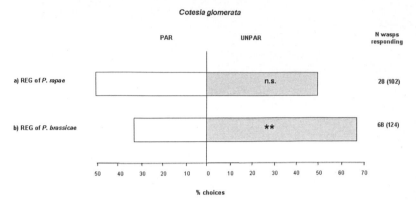

Figure 2: Response of *C. glomerata* to artificially damaged cabbage leaves treated with regurgitant (REG) from parasitized (PAR) or unparasitized (UNPAR) *Pieris spp.* caterpillars of a) 4th instar *P. rapae* larvae and b) 5th instar *P. brassicae* larvae, and in a two-choice set-up. Asterisks indicate significant differences within a choice test: ** $P < 0.01$, n.s. not significant (binomial test). The numbers show the number of wasps responding; the numbers in brackets show the total number of wasps tested. All wasps were released on a leaf damaged by first instar larvae either of UNPAR (a) *P. brassicae* or (b) *P. rapae*.

hosts in flight by using host-induced plant volatiles. Females of *C. glomerata* discriminated between host-induced plant volatiles induced by unparasitized and parasitized larvae in the case of their preferred host *P. brassicae*. As a result, the wasps save time compared to wasps that first have to land and explore a patch to detect non-volatile marking substances left by previous parasitoid visitors. Additionally, *C. glomerata* females reduce the risk of being bitten or getting in contact with the regurgitant of *P. brassicae* caterpillars. *Pieris brassicae* caterpillars, unlike *P. rapae* caterpillars, vigilantly defend themselves and can seriously harm their parasitoids (Brodeur et al. 1996).

Especially for a solitary endoparasitoid like *C. rubecula*, host discrimination is crucial because supernumerary larvae compete to the extent that only a single parasitoid can emerge (Salt 1961). The older larva usually eliminates the younger. The extent of the disadvantage depends critically on the time that elapses between the two ovipositions. Thus, in cases with a limited number of hosts present, parasitoids should be more willing to attack recently parasitized hosts in which their larvae have a higher probability to survive. Internal markers released by the egg or physiological changes in the host caused by parasitism become more noticeable with time (Godfray 1994 and references therein; van Lenteren 1976).

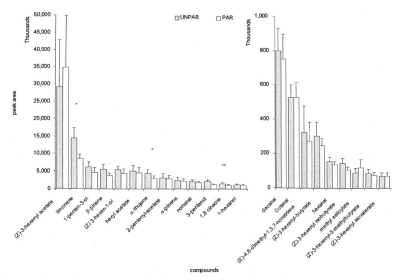

Figure 3: Comparison of the volatile blends emitted by cabbage plants artificially damaged and treated with regurgitant of unparasitized (UNPAR) or parasitized (PAR) caterpillars of *P. brassicae* for 22 compounds expressed as mean peak area (in arbitrary units) and the standard error. For each compound the mean of 9 peak areas of the UNPAR treatments and 9 peak areas of the PAR treatments are shown. Asterisks indicate significant differences: * $P < 0.05$; ** $P < 0.01$ (Wilcoxon's matched pairs signed rank test).

In gregarious parasitoids like *C. glomerata*, scramble competition may occur, i.e. no aggressive behavior towards other parasitoid larvae or physiological suppression occurs, and the parasitoids develop as long as food is available. Superparasitism can be an adaptive strategy under some conditions such as when wasps are faced with a limited number of hosts for a longer period. However, an increase in clutch size is often associated with a decrease in size of surviving offspring and a reduced fitness of the emerging adults (e.g. Harvey et al. 1993; Potting et al. 1997).

Naive females of *C. glomerata* have been shown to discriminate between parasitized and unparasitized hosts in close-range tests (Ikawa and Suzuki 1982; Le Masurier 1990). Experienced *C. glomerata* wasps were shown to regulate the number of eggs laid in response to the number of available suitable hosts (Ikawa and Suzuki 1982). In the present study we used females experienced with a particular plant-host-complex (as described before), to assess whether they discriminated in flight. Therefore, it remains unclear if the response was innate or triggered by the short odor experience just prior to the release in the windtunnel. Other studies showed that a single experience with a plant-host-complex had no influence on the behavior of either *C. glomerata* or *C. rubecula* towards the offered plant-host complexes in the windtunnel (Geervliet et al. 1998). However, in our case the brief experience resulted in a higher response of *C. glomerata* that flew to one of the odor sources in tests with plants treated with regurgitant of *P. brassicae*.

When offering leaves induced with regurgitant of *P. rapae*, a high proportion of females of our Dutch strain of *C. glomerata* were unresponsive, and those that did fly towards the plants did not discriminate. In the Netherlands, *C. glomerata* specializes on *P. brassicae* which has been ascribed to the fact that *C. rubecula* outcompetes *C. glomerata* in *P. rapae* larvae (Geervliet et al. 2000). The solitary endoparasitoid *C. rubecula* always wins the competition for the same host against the gregarious *C. glomerata* (Laing and Corrigan 1987). Our results suggest that the priming of *C. glomerata* wasps with leaf damage of first instar *P. rapae* caterpillars that we practiced just prior to the bioassay induced a low response and failure to discriminate between plants treated with regurgitant of unparasitized or parasitized *P. rapae* larvae.

To the best of our knowledge this is the first report in which elicitation of herbivore-induced plant volatile production, by controlling the amount of mechanical leaf damage combined with the application of regurgitant, was shown to result in a reduced emission of specific plant volatiles when elicitors originated from parasitized caterpillars. Souissi et al. (1998) reported that unparasitized mealybug-infested plants or unparasitized mealybugs were more attractive to walking *Apoanagyrus lopezi* (Hymenoptera: Encyrtidae) endoparasitoids than plants infested with parasitized mealybugs or parasitized mealybugs alone. The amount of herbivore feeding was not controlled in their study and it was not checked whether the plant was the source of the attractants. Sheehan et al. (1993) showed that females of *Microplitis croceipes* (Hymenoptera: Braconidae) were able to discriminate in flight between previously visited and unvisited sites in a plant patch even in the absence of the host. In this case, herbivore-induced plant volatiles were not tested, and it was assumed that a chemical marking deposited on the substrate served as the cue for discrimination.

At present it is still unknown for any parasitoid species exactly which compounds are involved as synomones (Dicke and van Loon 2000). Previous work on volatiles emitted by a different Brussels sprouts cultivar Blaakmeer et al. (1994) showed differences between intact and caterpillar-damaged plants for hexyl acetate, (Z)-3-hexenyl acetate, sabinene, and 1,8-cineole. These compounds were also increased in the headspace of plants treated with regurgitant of unparasitized *P. brassicae* caterpillars compared to treatments with regurgitant of parasitized *P. brassicae* caterpillars. The headspace composition documented here also overlaps with that described by Smid et al. (2002) who studied yet another Brussels sprouts cultivar and used feeding first instar caterpillars during headspace collection. An obvious difference is that benzyl cyanide, a breakdown product of glucosinolates, the taxonomically characteristic secondary plant metabolites of Brassicaceous plants, is lacking in our samples. It has recently been established (Wittstock et al. 2004) that this compound is released from the faeces of *Pieris* caterpillars, which explains its absence in our samples. Limonene elicits an olfactory response in the antennae of both *Cotesia* species (Smid et al., 2002). The fact that the reduced emission of three volatiles correlated with discriminatory behavior that were shown to elicit electrophysiological activity suggests that the quantity of these compounds, possibly relative to those of other compounds, provides information to the parasitoids. We have used leaves that were detached just prior to the experiments. For maize it has been shown that excision of leaves may influence the emission rate of certain components in some treatments at some times of the day, without affecting the overall effect of treatment (Schmelz et al. 2001). In our experiments, the treatments were maximally standardized with regard to leaf physiological age and size, and leaves from both treatments were excised to investigate locally emitted volatiles only.

Studies on the interrelationship of *Cotesia* larvae with their hosts showed drastic effects of parasitism on the caterpillars' physiology such as changes in hemolymph protein titer and composition (reviewed by Beckage 1993). Application of ß-glucosidase, a component of *P. brassicae* regurgitant, to artificially damaged plants resulted in the release of a volatile blend similar to that of leaves treated with regurgitant (Mattiacci et al. 1995). Quantitative differences in headspace composition resulting from treating plants with regurgitant obtained from either parasitized or unparasitized caterpillars might be due to a reduced amount of ß-glucosidase in regurgitant of parasitized *P. brassicae*.

Parasitization of *P. rapae* by *Cotesia rubecula* has been demonstrated to increase fitness of the brassicaceous plant *Arabidopsis thaliana* (van Loon et al. 2000). In the case of *C. glomerata*, which preferentially parasitizes the gregariously feeding *P. brassicae*, only a suppression of specific synomones, as documented here, would benefit the wasp as for this parasitoid-host combination an overall reduction in the amount of volatiles released could signal a lower but still sufficient number of available unparasitized hosts rather than parasitized hosts. As we controlled the amount of mechanical damage and the volume of elicitor-containing regurgitant, we can exclude the involvement of other stress factors, like viral, bacterial or fungal pathogens, that might have interfered with the feeding performance of the caterpillars and consequently with induced synomone emission by the plants.

The induced volatiles bear biosynthetic and ecological costs (Dicke and Sabelis 1989; Dicke and van Loon 2000; Hoballah et al. 2004). Many of the compounds are produced through *de novo* biosynthesis (Paré and Tumlinson 1997). Ecological costs in terms of attraction of herbivores can be much higher than the biosynthetic costs (Dicke and Vet 1999). Thus, down-regulation of induced volatile production is likely to be advantageous for plants.

In conclusion, we have shown that parasitoids can use herbivore-induced plant volatiles for host discrimination in flight, a crucial ability to enhance their reproductive success. Besides the fact that plant synomones are reliable and detectable indicators of the presence and identity of herbivores, we document that parasitoids can use them as a source of information on host suitability. Consequently, parasitoids can save energy and time in finding suitable hosts. From the plant's perspective our study uncovered an advantage through a reduced production of a subset of herbivore-induced volatiles, after having successfully recruited their bodyguards.

Acknowledgements

We thank Joachim Ruther, Joop van Lenteren, and Monika Hilker for their advice on this research and valuable comments on an earlier version of this manuscript; Leo Koopman, Frans van Aggelen, and André Gidding for culturing the insects. Funding by the Bresillac Foundation is gratefully acknowledged. All experiments complied with the current laws of The Netherlands.

References

Adams RP (2001) Identification of essential oil components by gas chromatography/mass spectroscopy. Allured Publishing Corporation, Carol Stream, Illinois

Agelopoulos NG, Keller MA (1994) Plant-natural enemy association in the tritrophic system, *Cotesia rubecula-Pieris rapae*-Brassicaceae (Cruciferae): II. Preference of *C. rubecula* for landing and searching. Journal of Chemical Ecology 20:1735-1748

Beckage NE (1993) Games parasites play: The dynamic roles of proteins and peptides in the relationship between parasite and host. In: Beckage NE, Thompson SN, Federici BA (eds) Parasites and pathogens of insects, vol 1. Academic Press, San Diego, pp 25-57

Blaakmeer A, Geervliet JBF, van Loon JJA, Posthumus MA, van Beek TA, de Groot Æ (1994) Comparative headspace analysis of cabbage plants damaged by two species of *Pieris* caterpillars: consequences for in-flight host location by *Cotesia* parasitoids. Entomologia Experimentalis et Applicata 73:175-182

Brodeur J, Geervliet JBF, Vet LEM (1996) The role of host species, age and defensive behaviour on ovipositional decisions in a solitary specialist and gregarious generalist parasitoid (*Cotesia* species). Entomologia Experimentalis et Applicata 81:132

Dicke M (1999) Are herbivore-induced plant volatiles reliable indicators of herbivore identity to foraging carnivorous arthropods? Entomologia Experimentalis et Applicata 91:131-142

Dicke M, Sabelis MW (1989) Does it pay plants to advertize for bodyguards? Towards a cost-benefit analysis of induced synomone production. In: Lambers H, Cambridge ML, Konings H, Pons TL (eds) Causes & Consequences of Variation in Growth Rate and Productivity of Higher Plants. SPB Publishing, The Hague, pp 341-358

Dicke M, van Loon JJA (2000) Multitrophic effects of herbivore-induced plant volatiles in an evolutionary context. Entomologia Experimentalis et Applicata 97:237-249

Dicke M, Vet LEM (1999) Plant-carnivore interactions: evolutionary and ecological consequences for plant, herbivore and carnivore. In: Olff H, Brown VK, Drent RH (eds) Herbivores: Between Plants and Predators. Blackwell Science, Oxford, pp 483-520

Geervliet J, B.F., Verdel MSW, Henk S, Schaub J, Dicke M, Vet LEM (2000) Coexistence and niche segregation by field populations of the parasitoids *Cotesia glomerata* and *C. rubecula* in the Netherlands: predicting field performance from laboratory data. Oecologia 124:55-63

Geervliet JBF, Vet LEM, Dicke M (1994) Volatiles from damaged plants as major cues in long-rage host-searching by the specialist parasitoid *Cotesia rubescula*. Entomologia Experimentalis et Applicata 73:289-297

Geervliet JBF, Vreugdenhil AI, Dicke M, Vet LEM (1998) Learning to discriminate between infochemicals from different plant-host complexes by the parasitoids *Cotesia glomerata* and *C. rubecula*. Entomologia Experimentalis et Applicata 86:241-252

Godfray HCJ (1994) Parasitoids - Behavioral and Evolutionary Ecology. Princeton University Press, Princeton, New Jersey

Harvey JA, Harvey IF, Thompson DJ (1993) The effect of superparasitism on development of the solitary wasp parasitiod *Venturia canescens* (Hymenoptera: Ichneumonidae). Ecological Entomology 18:203-208

Hilker M, Meiners T (2002) Induction of plant responses towards oviposition and feeding of herbivorous arthropods: a comparison. Entomologia Experimentalis et Applicata 104:181-192

Hoballah ME, Kollner TG, Degenhardt J, Turlings TCJ (2004) Costs of induced volatile production in maize. Oikos 105:168-180

Ikawa T, Suzuki Y (1982) Ovipositional experience of the gregarious parasitoid, *Apanteles glomeratus* (Hymenoptera: Braconidae), influencing her discrimination of the host larvae, *Pieris rapae crucivora*. Applied Entomology and Zoology 17:119-126

Laing JE, Corrigan JE (1987) Intrinsic competition between the gregarious parasite, *Cotesia glomeratus* and the solitary parasite, *Cotesia rubecula* (Hymenoptera: Braconidae) for their host, *Artogeia rapae* (Lepidoptera: Pieridae). Entomophaga 32:493-501

Le Masurier AD (1990) Host discrimination by *Cotesia* (=*Apanteles*) *glomerata* parasitising *Pieris rapae*. Entomologia Experimentalis et Applicata 54:65-72

Mattiacci L, Dicke M, Posthumus MA (1994) Induction of parasitoid attracting synomone in Brussels sprouts plants by feeding of *Pieris brassicae* larvae: role of mechanical damage and herbivore elicitor. Journal of Chemical Ecology 20:2229-2247

Mattiacci L, Dicke M, Posthumus MA (1995) ß-Glucosidase: An elicitor of the herbivore-induced plant odor that attracts host-searching parasitic wasps. Proceedings of the National Academy of Sciences of the United States of America 92:2036-2040

Mattiacci L, Rudelli S, Ambühl Rocca B, Genini S, Dorn S (2001) Systemically-induced response of cabbage plants against a specialist herbivore, *Pieris brassicae*. Chemoecology 11:167-173

Nufio CR, Papaj DR (2001) Host marking behavior in phytophagous insects and parasitoids. Entomologia Experimentalis et Applicata 99:273-293

Paré PW, Tumlinson JH (1997) Induced synthesis of plant volatiles. Nature 385:30-31

Potting RPJ, Snellen HM, Vet LEM (1997) Fitness consequences of superparasitism and mechanism of host discrimination in the stemborer parasitoid *Cotesia flavipes*. Entomologia Experimentalis et Applicata 82:341-348

Ruther J (2000) Retention index database for identification of general green leaf volatiles in plants by coupled capillary gas chromatography-mass spectrometry. Journal of Chromatography A 890:313-319

Salt G (1961) Competition among insect parasitoids. Symp. Soc. Exp. Biol. 15:96-119

Schmelz EA, Alborn HT, Tumlinson JH (2001) The influence of intact-plant and excised-leaf bioassay designs of volicitin- and jasmonic acid-induced sesquiterpene volatile release in *Zea mays*. Planta 214:171-179

Sheehan W, Wäckers FL, Lewis WJ (1993) Discrimination of previously searched, host-free sites by *Micoplitis croceipes* (Hymenoptera: Braconidae). J. Insect Behav. 6:323-331

Shiojiri K, Takabayashi J, Yano S, Takafuji A (2000) Herbivore-species-specific interactions between crucifer plants and parasitic wasps (Hymenoptera : Braconidae) that are mediated by infochemicals present in areas damaged by herbivores. Applied Entomology and Zoology 35:519-524

Shiojiri K, Takabayashi J, Yano S, Takafuji A (2001) Infochemically mediated tritrophic interaction webs on cabbage plants. Popul.Ecol. 43:23-29

Siegel S (1956) Nonparametric Statistics for the Behavioral Sciences. McGraw-Hill Kogakusha, Ltd., Tokyo

Smid HM, Van Loon JJA, Posthumus M, A., Vet LEM (2002) GC-EAG-analysis of volatiles from Brussels sprouts plants damaged by two species of *Pieris* caterpillars: olfactory respective range of a specialist and generalist parasitoid wasp species. Chemoecology 12:169-176

Souissi R, Nenon J-P, Le Rü B (1998) Olfactory responses of parasitoid *Apoanagyrus lopezi* to odor of plants, mealybugs, and plant-mealybug complexes. Journal of Chemical Ecology 24:37-48

Steidle JLM, van Loon JJA (2003) Dietary specialization and infochemical use in carnivorous arthropods: testing a concept. Entomologia Experimentalis et Applicata 108:133-148

Takabayashi J, Dicke M (1996) Plant-carnivore mutualism through herbivore-induced carnivore attractants. Trends in plant science 1:109-113

Takasu K, Hirose Y (1988) Host discrimination in the parasitoid *Ooencyrtus nezarae*: the role the egg stalk as an external marker. Entomologia Experimentalis et Applicata 47:47-48

Turlings TCJ, Gouinguené S, Degen T, Fritzsche Hoballah ME (2002) The chemical ecology of plant-caterpillar-parasitoid interactions. In: Tscharntke T, Hawkins BA (eds) Multitrophic Level Interactions. Cambridge University press, Cambridge, pp 148-173

Turlings TCJ, Loughrin JH, McCall PJ, Röse USR, Lewis WJ, Tumlinson JH (1995) How caterpillar-damaged plants protect themselves by attracting parasitic wasps. PNAS USA 92:4169-4174

van Alphen JJM, Visser ME (1990) Superparasitism as an adaptive strategy for insect parasitoids. Annual Review of Entomology 35:59-79

van Baaren J, Nenon J-P (1996) Host location and discrimination mediated through olfactory stimuli in two species of Encyrtidae. Entomologia Experimentalis et Applicata 81:61-69

van Giessen WA, Lewis WJ, Vet LEM, Wäckers FL (1993) The influence of host site experience on subsequent flight behavior in *Microplitis croceipes* (Cresson) (Hymenoptera: Braconidae). Biological Control 3:75-79

van Lenteren JC (1976) The development of host discrimination and the prevention of superparasitism in the parasite *Pseudeucoila bochei* (Hym.: Cynipidae). Netherlands Journal of Zoology 26:1-83

van Lenteren JC (1981) Host discrimination by parasitoids. In: Nordlund DA, Jones RL, Lewis WJ (eds) Semiochemicals- Their Role in Pest Control. John Wiley & Sons, New York, pp 153-179 van Loon JJA, de Boer JG, Dicke M (2000) Parasitoid-plant mutualism: parasitoid attack of herbivore increases plant reproduction. Entomologia Experimentalis et Applicata 97:219-227

Vet LEM, Dicke M (1992) Ecology of infochemical use by natural enemies in a tritrophic context. Annual Review of Entomology 37:141-172

Vinson SB (1985) The behavior of parasitoids. In: Kerkut GA, Gilbert LI (eds) Comprehensive Insect Physiology, Biochemistry and Pharmacology, vol 9. Pergamon Press, New York, USA, pp 417-469

Wittstock U et al. (2004) Successful herbivore attack due to metabolic diversion of a plant chemical defense. Proceedings of the National Academy of Sciences of the United States of America 101:4859-4864

Summarizing Discussion

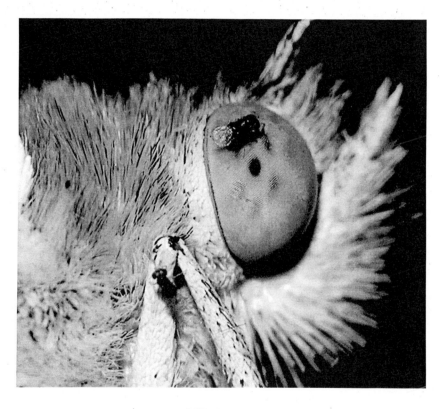

N.E. Fatouros

Chapter 9

Summarizing Discussion

Chemical cues from the host, host's food plant, or host environment are considered to be crucial for the parasitoid's foraging success. Parasitoids that search for herbivorous hosts face a reliability-detectability problem in their use of such infochemicals (Vet and Dicke 1992): due to eavesdropping enemies or production costs their hosts are under selection to emit chemical signals in the lowest quantities possible, which results in a low detectability. On the other hand, their host's food plants produce large amounts of detectable chemical signals, albeit of low reliability for indicating the host's presence.

This thesis provides examples of solutions to this host location dilemma for two different types of parasitic wasps: *Trichogramma* and *Cotesia* parasitoids "hunting" for the eggs and larvae of the same host – the large cabbage white *Pieris brassicae* (Lepidoptera: Pieridae). The first part of this thesis reveals two novel host finding strategies of *Trichogramma brassicae* to locate *P. brassicae* eggs by a) using the chemical espionage-and-ride strategy and b) exploiting oviposition-induced plant synomones. The second part of the thesis elucidates a novel component of specificity of larval-induced plant volatiles and its exploitation by two conspecific *Cotesia* parasitoids during the host acceptance phase. I addressed the following questions:

(1) Do *Trichogramma* egg parasitoids exploit the communication system of *P. brassicae* butterflies?

(2) Do *Trichogramma* egg parasitoids make use of synomones of Brussels sprouts plants induced by *P. brassicae* eggs? These new aspects allow comparisons to be made to the three other systems, where indirect plant defense by herbivorous eggs was demonstrated and to the already shown induction induced by larval feeding (See table 1, compare chapter 1).

(3) Do *Cotesia* larval parasitoids make use of volatiles induced in Brussels sprouts plants by host feeding to discriminate between suitable and unsuitable, already parasitised hosts in-flight?

This discussion intends to synthesize the main results and conclusions of the first part of the thesis, i.e. the research on *Trichogramma* wasps, with the second part on *Cotesia* host discrimination and compares them with general patterns of different strategies of larval and egg parasitoids. Furthermore, the functions and mechanisms of herbivore-induced plant synomones will be discussed here and set in a broader ecological perspective.

Infochemical-exploiting strategies used by egg and larval parasitoids of herbivorous insects

Functions of host kairomones

Infochemicals of the host are reliable cues for foraging parasitoids (Vet and Dicke 1992). However, their hosts are under strong selection pressure to remain as inconspicuous as possible to avoid being detected by their enemies. They can do so by, e.g., choosing plants or plant parts for feeding or oviposition not easy to find by their enemies (enemy free space) (e.g. Berdegue et al. 1996; Denno et al. 1990; Gratton and Welter 1999; Gross et al. 2004). Yet, herbivorous host insects can not completely preclude essential behaviors such as intraspecific communication, frass production or defense all releasing infochemicals that reveal the presence of the producer. In chapter 2, I reviewed the different infochemical-exploiting strategies used by egg parasitoids. Egg parasitoids strongly depend on cues of the adult host stage, since their host eggs hardly emit any long range cues. Sexual pheromones produced by lepidopteran hosts are frequently used by egg parasitoids to find oviposition sites or adult hosts for transportation to such oviposition sites (i.e., phoresy) (e.g. Arakaki et al. 1996; Fatouros et al. 2005b; Powell 1999) (Chapter 3).

Few parasitoids of larval instars have been shown to spy on sexual host pheromones because the juvenile stage attacked is less closely associated with the adult insect than eggs are. Still, this strategy can be advantageous to larval parasitoids when the pheromone-releasing adult and the larvae appear in the same place at the same time. The synthetic sex pheromones of some moth females attracted the braconid wasps *Cotesia ruficrus* and *Microplitis rufiventris* in olfactory tests (Zaki 1996). Cortesero et al. (1993) observed an increased locomotory activity towards the odor of sexually active female bruchid beetles but found no evidence of attraction. In bruchid beetles adult and larval hosts often coexist. Some aphid parasitoids were shown to use the synthetic compounds of sex pheromones released by sexual host females though evidence on parasitoids using naturally produced aphid sex pheromones for location in the field was not given (Gabrys et al. 1997; Glinwood et al. 1999; Powell 1999; Powell et al. 1998). Aphid parasitoids are known to oviposit in larval stages as well as in adult stages and both of them often coexist.

Some larval parasitoids spy on different types of pheromones such as aggregation pheromones of adult bark beetles or fruit flies (Kennedy 1984; Wiskerke et al. 1993), oviposition-deterring pheromones of tephritid flies (Prokopy and Webster 1978), other so-called epideictic pheromones (reviewed by Powell 1999), or aphid alarm pheromones (Grasswitz and Paine 1992; Powell 1999).

Numerous other host cues can be of importance to arrest or attract egg and larval parasitoids. Egg parasitoids mainly use close-range cues again from the adult host such as wing scales of lepidopteran hosts (Fatouros et al. 2005a; Jones et al. 1973; Laing 1937; Lewis et al. 1975; Lewis et al. 1972; Noldus and van Lenteren 1985; Shu and Jones 1989; Zaborski et al. 1987) (Chapter 4 and 7), or excretions of female accessory glands (Nordlund et al. 1987; Strand and Vinson 1982) (Chapter 6). In contrast, larval parasitoids can use cues of the attacked host stage itself. Important sources can be e.g., chemicals produced by larval frass or mandibular/labial gland secretions (reviewed by Godfray 1994; Rutledge 1996), or feces (e.g. Agelopoulos et al. 1995; Cortesero et al. 1993; Geervliet et al. 1994; Turlings et al. 1991).

In chapter 3 we demonstrated that *T. brassicae* egg parasitoids respond to the anti-aphrodisiac pheromone of *P. brassicae* females, benzyl cyanide. *T. brassicae* first mounts a mated female butterfly marked with the anti-aphrodisiac and thereafter hitch-hikes with her to a host plant. When the *Pieris* female starts to lay eggs, the wasp descends onto the leaf and parasitises the females' freshly laid eggs. Thus, the anti-aphrodisiac leads the egg parasitoid faster to its target, the host eggs, unlike with sex pheromones, which are emitted by females that first have to mate before they oviposit. So far, anti-aphrodisiacs have been identified in *Drosophila melanogaster* (Scott 1986), meal worms (Happ 1969), bees (Frankie et al. 1980; Kukuk 1985), and *Heliconius* and *Pieris* butterflies (Andersson et al. 2000; 2003; Gilbert 1976) (See also figure 1). Based on the rising number of insect species identified to use anti-aphrodisiacs, it can be expected that egg parasitoids use this strategy of chemical espionage on host anti-aphrodisiacs more frequently in nature. However, it is doubtful if parasitoids of larval stages, such as *Cotesia* spp., make use of *Pieris* anti-aphrodisiacs. Several days pass from the release of the pheromone of mated *Pieris* females until the hatch of the larvae. Moreover, relying on the anti-aphrodisiac would not guarantee *Cotesia* wasps to find oviposition sites but rather the adult host. Unlike egg parasitoids, larval parasitoids have not been reported so far to execute phoretic behavior on adult hosts. Yet, it can not be excluded that the pheromone is released during egg deposition by *Pieris* spp. and is then detectable the first few hours for foraging *Cotesia* parasitoids. The egg parasitoid *T. brassicae* did not show any arrestment to Brussels sprouts plants treated with synthetic benzyl cyanide tested 24 h after treatment (Chapter 6). Here, an attraction to the anti-aphrodisiac right after application on the leaf was not tested. Generally, it can be concluded that egg parasitoids rely much more on pheromones, mainly those functioning during mate finding, and close-range cues of the adult host stage than larval parasitoids do.

Functions of plant synomones

Plants release volatile blends upon herbivore feeding that can be exploited by natural enemies during host or prey location (e.g. Dicke and Sabelis 1988; Turlings et al. 1990). Turlings et al. (1990) were the first to show an attraction of larval

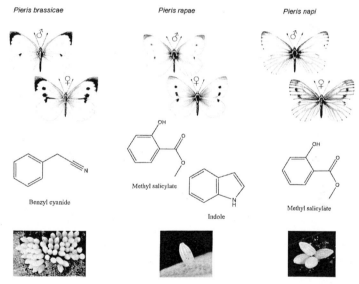

Figure 1: Variation in anti-aphrodisiac production in three different *Pieris* spp. (Figure drawn by M.E. Huigens)

parasitoids to volatiles from herbivore-infested plants. More studies followed with parasitoids of caterpillars on cotton (Loughrin et al. 1994; McCall et al. 1993; Röse et al. 1996), cabbage (Agelopoulos and Keller 1994; Geervliet et al. 1994; Mattiacci et al. 1994; Steinberg et al. 1992), and maize (Potting et al. 1995; Takabayashi et al. 1995; Turlings et al. 1991), all demonstrating that caterpillar feeding damage releases plant synomones that attract parasitoids. A specificity of such synomonal cues was demonstrated for several systems and aspects. For example, it was shown that volatiles from corn plants induced by host instars suitable for parasitization attracted more *Cotesia* parasitoids than conspecific plants infested with less suitable host instars (Takabayashi et al. 1995). However, in other tritrophic systems this was not the case (Gouinguené et al. 2003; Mattiacci and Dicke 1995b). Different herbivore species can cause distinctive responses in a plant resulting in qualitatively and/or quantitatively differing blend compositions that are discriminated by the parasitoids (Du et al. 1998; Turlings et al. 1998; Turlings et al. 1993). Furthermore, some studies revealed that plants can even "choose" what and when to emit (De Moraes et al. 2001; Kahl et al. 2000). For example, tobacco plants emit different blends after caterpillar feeding between day and night. During the day they attract parasitoids, whereas they repel the nocturnally active herbivores from egg deposition during night time (De Moraes et al. 1998; 2001).

In this thesis, I present a novel function of herbivore-induced plant volatiles demonstrated for the two parasitoids, *Cotesia glomerata* and *C. rubecula*. It was shown that the wasps discriminated between odors from Brussels sprouts plants treated with unparasitized caterpillars or their regurgitant and plants treated with parasitized caterpillars or their regurgitant of their specific hosts (either *P. brassicae* and/or *P. rapae*) in flight (Fatouros et al. 2005c) (Chapter 8). Such a host discrimination behavior by parasitoids was so far known to be mediated by marking pheromones, physiological changes in the host or of the host surface usually perceived through contact (Nufio and Papaj 2001; Takasu and Hirose 1988; van Lenteren 1981). Furthermore, I demonstrated that Brussels sprouts plants treated with the regurgitant of unparasitized or parasitized caterpillars differed in the total quantity of volatiles released. Hereby, three major terpenoids from plants treated with regurgitant of parasitized caterpillars were emitted in lower amounts. A downregulation of the induced synomone emission can be advantageous for the plant since volatile production is "expensive": biosynthetic and ecological costs need to be "paid" (Dicke 1999; Dicke et al. 2003; Hoballah et al. 2004). The results of chapter 8 reveal how sophisticated plants are in their response towards herbivore infestation and how well adapted the response of some natural enemies to this specificity can be. Still, due to variations in the composition of the blend between different plant genotypes, plant parts, growth stages etc. such signals can be limited in their specific information they provide to natural enemies (reviewed by Dicke et al. 2003; Turlings and Wäckers 2004).

Also egg parasitoids were shown to respond to volatiles induced by feeding (Lou et al. 2005; Moraes et al. 2005). Responses to feeding-induced volatiles can be adaptive to egg parasitoids when all developmental host stages co-occur on the same plant, which makes an association of host adult or nymph feeding with egg presence a reliable cue. Plants infested by young *P. brassicae* caterpillars did not arrest *T. brassicae* wasps, regardless whether eggs were additionally present or not (Fatouros et al. 2005a) (See also figure 2). *Pieris* caterpillars normally hatch quite simultaneously from one egg clutch and first start feeding on the egg rests. It is therefore not very likely that feeding co-occurs with host eggs suitable for parasitization. Additionally, the adult butterflies themselves are nectar feeders and usually are deterred by plants laden with egg clutches (Klijnstra 1986; Rothschild and Schoonhoven 1977).

Recently, egg parasitoids were shown to be attracted to plants induced by oviposition by herbivorous insects (reviewed by Hilker and Meiners 2002; 2006). Evidence for oviposition-induced plant synomones attracting egg parasitoids exists up to now from four studied tritrophic systems named after the plant: the elm system (Meiners and Hilker 1997; Meiners and Hilker 2000; Wegener et al. 2001), pine system (Hilker et al. 2002; Mumm et al. 2003), bean system (Colazza et al. 2004a; b), and cabbage system (Fatouros et al. 2005a) (Chapter 4-6). In the cabbage system, I have demonstrated an interaction between *T. brassicae* egg parasitoids using cues of the Brussels sprouts surfaces induced by *P. brassicae* eggs (Fatouros et al. 2005a) (Chapter 4, 5 & 6), which shows a novel indirect plant defense mechanism so far not documented. I found an indication for a reduction of the glucobrassicin derivate, indole-3-acetonitrile, in the leaf surfaces of egg-infested plants (Chapter 5). Although behavioral proof has not been obtained yet, I hypothesize that these modifications in the glucosinolate profile of the leaf surface by egg deposition might serve as chemical cues for the egg parasitoid *T. brassicae*. Glucosinolates are deterrent to many herbivorous insects. Higher concentrations in uninfested plants could as well repel foraging parasitic wasps, whereas a reduction after egg deposition could arrest them and through learning they could associate such induced plants with egg presence.

So far, larval parasitoids have not been shown to respond to volatiles of egg-infested plants. *C. glomerata* wasps were tested whether they exploit such an infochemical detour strategy. Naïve wasps did not prefer odors of a plant with *P. brassicae* eggs, seven days after deposition or else one day before larval hatching (Mattiacci and Dicke 1995a). Still, preliminary bioassays showed an increased searching and probing behavior of the *Cotesia* wasps among the eggs deposited on a leaf. A response to odors from egg-infested plants could be advantageous for the larval parasitoids when the eggs are close to hatching. *C. glomerata* did not discriminate between plant volatiles induced by different old larval host stages as was shown for another *Cotesia* spp. (Takabayashi et al. 1995). By using odors

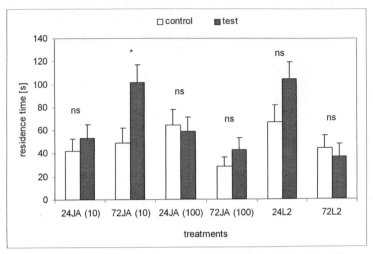

Figure 2: Arrestment of *T. brassicae* to plants treated with JA (10 and 100 μM) or 2nd instars (L2) of *P. brassicae* 24 or 72h after treatment. Number of tested wasps per treatment N=30-45. Wilcoxon's matched pairs signed rank test, * $p<0.05$, n.s. not significant.

of egg-laden plants they could arrive in time for the favorable first instars since contacts with older instars can be risky because of their defensive behavior. It could be possible that only parasitoid females with an oviposition experience in just emerging caterpillars are able to learn such an infochemical detour. On the other hand, the results of the behavioral tests with *T. brassicae* gave no indications for an induction of Brussels sprouts volatiles in response to *P. brassicae* egg depositions. Yet, a response to chemical modifications in the leaf surface after egg deposition could as well help the *Cotesia* wasps to find newly hatching caterpillars.

Generally, an attraction/arrestment to herbivore-induced plant synomones was shown to be a solution to the reliability-detectability problem (Compare chapter 2) for both larval and egg parasitoids. Egg parasitoids are able to use either synomones induced by the eggs themselves or by feeding host stages. Both types of plant cues can be very specific in indicating the presence of host eggs, and thus might play equally important roles for foraging egg parasitoids. However, an attraction of larval parasitoids to oviposition-induced synomones could be a reliable cue as well, e.g., in those cases when feeding-induced synomones are not specifically indicating the presence of suitable young host stages to the parasitoids.

Mechanisms of induction of plant synomones
Oral secretions and regurgitants of herbivorous insects are known to elicit

indirect induced plant responses (e.g. Alborn et al. 1997; Mattiacci et al. 1995; Pohnert et al. 1999). Applications of compounds such as the enzyme β-glucosidase or fatty-acid-amino-acid conjugates like volicitin onto wounded plants were shown to induce the emission of a volatile blend attractive to parasitoids. Plasma membrane proteins in maize were suggested to act as a binding protein for volicitin (Truitt et al. 2004). Elicitors of oviposition-induced plant responses were found in the oviduct secretions of the elm leaf beetle and the pine sawfly (Hilker et al. 2005; Meiners and Hilker 2000). The elicitor of the pine sawfly was suggested to be a small protein.

In Chapter 6, I demonstrated that the eliciting activity of *P. brassicae* eggs is in the secretion associated with the eggs, i.e., secretion of the accessory reproductive glands (ARG) of mated female *P. brassicae*. In contrast, ARG secretion obtained from virgin females had no activity. During mating the *P. brassicae* male transfers the anti-aphrodisiac, benzyl cyanide (BC) within the spermatophore into the bursa copulatrix of the female (Andersson et al. 2003). Benzyl cyanide applied in a higher dosage or in traces in a mixture together with the ARG secretion from virgin females induced leaf surface changes that significantly arrested the wasps three days after application. It seems that BC is either activating the eliciting function of the ARG secretion or is just better soluble in the ARG secretion and therefore then functioning also in smaller doses. The polar BC could be transported through polar pathways through the epidermis or bound to plasma-membrane proteins to induce the signaling pathway required for an herbivore-induced response.

Like elicitors detected so far in oviposition-induced plant responses, most elicitors of feeding-induced responses have more polar characteristics. β-Glucosidase was found in the regurgitant of *P. brassicae* caterpillars. Application of this, in insects and plants very common, enzyme on mechanically-wounded Brussels sprouts plants elicited a volatile blend similar to that of regurgitant treated leaves (Mattiacci et al. 1995). The enzyme is produced by the caterpillar or associated microorganisms most likely in the mouth area. In chapter 8, I showed that the regurgitant of parasitized *P. brassicae* caterpillars applied on Brussels sprouts leaves downregulated the induction of *Cotesia*-attracting plant volatiles (Fatouros et al. 2005c). In enzyme activity analyses, the activity of β-glucosidase in the regurgitant of parasitized and unparasitized *P. brassicae* last instars was tested. There was an indication for a decrease of the enzyme activity level in the regurgitant of parasitized caterpillars; however, future studies have to examine this (Fatouros, unpublished results). Koinobiont parasitoids are known to induce physiological, biochemical, and behavioral changes in their hosts (Beckage 1993).

The total amount of hemolymph protein in *Manduca sexta* larvae parasitized by *C. congregata* was reduced almost immediately after parasitism. Possible inhibition factors were hormonal modulators of host protein synthesis, or regulation of host gene expression by factors associated with the polydnavirus or molecules secreted by the parasitoids (Beckage and Kanost 1993). It has to be tested whether the *C. glomerata* larvae inside the *P. brassicae* caterpillars influence the enzyme titer through a downregulation of the expression of the β-glucosidase gene.

With regard to the sort of volatile compounds produced by herbivorous feeding, terpenoids seems to play a major role in the volatile blend. Concerning plants induced by feeding herbivores, several studies indicated increased amounts or *de novo* synthesized sesquiterpenes after feeding, regurgitant treatment, elicitor or jasmonic acid treatment (e.g. Gols et al. 1999; Paré and Tumlinson 1997; Schmelz et al. 2001; Turlings et al. 1993). Two homoterpenes, 4,8-dimethyl-1,3(*E*),7-nonatriene (DMNT) and 4,8,12-trimethyl-1,3(*E*),7(*E*),11-tridecatetraene (TMTT), are often induced after herbivore feeding (reviewed by Dicke 1994). Terpenoids can be produced through the mevalonic acid pathway or the 1-deoxy-D-xylulose-5-phosphate pathway (reviewed by Dicke and van Poecke 2002). In Brussels sprouts plants, terpenoids seem to be increased in caterpillar-damaged plants as well (Blaakmeer et al. 1994; Mattiacci et al. 1994). Limonene was shown to elicit an olfactory response in the antennae of both *C. glomerata* and *C. rubecula* (Smid et al. 2002). After parasitization, a reduction in three terpenoids was measured in Brussels sprouts induced by regurgitant of *P. brassicae* caterpillars (Fatouros et al. 2005c)(Chapter 8).

Also after egg deposition plants seem to quantitatively change the emission of terpenoids that are detected by egg parasitoids (Colazza et al. 2004b; Mumm et al. 2003; Wegener and Schulz 2002; Wegener et al. 2001). In the pine and bean system, specific terpenoids were shown to be attractive to egg parasitoids such as (*E*)-β-farnesene within the volatile profile of egg-laden pine twigs or (*E*)-β-caryophyllene in extracts from host feeding damaged bean plants with egg batches (Colazza et al. 2004b; Mumm and Hilker 2005; Mumm et al. 2003). In Brussels sprouts plants induced by *P. brassicae* egg depositions no terpenes were found in leaf surface washes of egg-laden Brussels sprouts plants (Chapter 5). This difference to the other tritrophic systems where egg depositions induce plant volatiles or to plant responses to feeding might have one main cause: the lack of severe damage to the plant surface. *P. brassicae* females only slightly damage the upper wax layer during oviposition with their legs and/or the ovipositor (Figure 3).

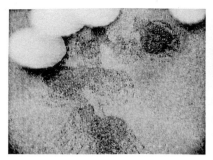

Figure 3: Slight damage of the epicuticular wax layer by scratches of the legs and/or ovipositor of *P. brassicae* females during oviposition on Brussels sprouts plants.

The plant hormone jasmonic acid (JA) has a central role in the induction of plant volatiles either by feeding or egg deposition: JA-treated plants result in the attraction of predators and parasitoids, as shown in several studies under laboratory and field conditions (Dicke et al. 1999; Gols et al. 1999; Hilker et al. 2002; Meiners and Hilker 2000; Ozawa et al. 2004; Thaler 1999). In the elm and pine system, JA-treated plants attract the egg parasitoids and chemical analysis revealed similar volatile profiles from JA-treated plants and egg-laden plants (Hilker et al. 2002; Meiners and Hilker 2000; Mumm et al. 2003; Wegener et al. 2001). In Brussels sprouts, JA treatment activates the lipoxygenase gene (*LOX*) (Zheng et al. 2006) that codes for the LOX enzyme, which mediates one of the steps in the green-leaf volatile production pathway (reviewed by Dicke and van Poecke 2002). In chapter 5, I tested the *LOX* gene activity in egg-laden Brussels sprouts plants; however, I did not detect any expression of the gene. Bioassays with *T. brassicae* wasps and JA-treated plants revealed that the egg parasitoids are arrested to leaves treated with 0.05 µg JA/cm2 JA 72 h after application (M. Bruinsma and N. Fatouros, unpublished results) (Figure 3). No arrestment of *T. brassicae* to JA-treated plants 24 h after application or to larval feeding was found. These behavioral results may be an indication for the involvement of JA in the induction of leaf surface changes that affect the behavior of the egg parasitoids. Chemical analysis of glucosinolates of the Brussels sprouts leaf surface indicated a decrease of an indole glucosinolates after JA treatment (M. Bruinsma, unpublished results). These results correspond to the chemical analysis of leaf surfaces from egg-laden plants, where indole-3-acetonitrile was reduced (see above) (Chapter 5). This could be an indication of glucosinolate contents or composition being involved in indirect defense responses of Brussels sprouts plants.

Interestingly, the results presented in this thesis show that different host stages of *P. brassicae* induce very different defense mechanisms in Brussels sprouts and consequently induce diverse responses in the egg and larval parasitoids (Compare Table 1). The signal cascade in the plant is evoked by two different elicitors, the nitrile benzyl cyanide within the accessory gland secretion on one hand, the enzyme β-glucosidase in combination with wounding on the other hand. Benzyl cyanide in combination with products associated with egg depositions leads to changes in the glucosinolate profile on and in the leaves, only detected through contact by *T. brassicae*. In contrast, larval feeding or β-glucosidase induces probably the lipoxygenase pathway and mevalonic acid pathway mediating green leaf volatile and terpenoid production attracting *Cotesia* parasitoids in flight. The egg parasitoid responded very specific to the synomone induced by *P. brassicae* eggs, but did not detect the synomone induced by *P. brassicae* caterpillars. Conversely, the larval parasitoids were shown to respond specifically to feeding-induced volatiles. Future studies need to evaluate whether experienced *Cotesia* spp. are also able to detect the synomone induced by egg deposition.

Table 1: Comparison of feeding and oviposition induced indirect plant responses in cabbage and in three tritrophic systems with oviposition induced plant responses

Mechanism/ plant	Induction by oviposition		Induction by feeding
	trees/bean[1]	Cabbage[2]	Cabbage[3]
Volatile blend	- mainly terpenoids	- **No!** Most likely close range cues in leaf surface	- about 20 important compound, e.g. terpenoids
	- By egg deposition + wounding/ feeding	- By egg deposition, no obvious wounding	- By larval feeding or regurgitant
Local	✔	✔	✔
Systemic	✔	✔	✔
Induction time	3 h – 5 d / 3 – 4 d	3 – 4 d	ca. 6 h
Elicitor	- In oviduct secretion	- In accessory gland secretion	- In regurgitant
	- Protein involved	- Benzyl cyanide involved	– β-glucosidase involved

[1] Colazza et al. 2004a; b; Hilker et al. 2002; Meiners and Hilker 1997; 2000; Mumm et al. 2003; Wegener et al. 2001

[2] Fatouros et al. 2005a; Chapters 4-6

[3] Blaakmeer et al. 1994; Fatouros et al. 2005b; Mattiacci et al. 1995; 2001; Smid et al. 2002

Conclusion and future perspectives

This thesis gives new insights into host foraging strategies of parasitoids exploiting infochemicals of Brussels sprouts plants and *Pieris* hosts. With respect to *Trichogramma* egg parasitoids, a complete sequence from habitat to host location was shown by the use of kairomones from the host as long-range cues to synomones from the host plant as short-range cues. Both, the kairomone and the synomone are related to a single compound, the anti-aphrodisiac pheromone benzyl cyanide, donated by the male *P. brassicae* butterflies to the females during mating. This anti-aphrodisiac of mated *P. brassicae* females lures *T. brassicae* wasps and by hitching a ride on them, they can reach their desired host eggs. Eggs laid by the host females induce a plant response arresting subsequent *T. brassicae* three days later. Again the male-contributed benzyl cyanide plays a central role: together with compounds of the accessory glands it is delivered onto the plant surface and induces the indirect plant defense mechanism. An additional novel indirect plant mechanism was shown to help *Cotesia* parasitoids to "decide" whether the hosts are suitable for parasitisation or not: by reducing the amount of feeding-induced volatiles after successful recruitment of the parasitoids, plants may save biosynthetic and ecological costs. Futhermore, *Cotesia* parasitoids save energy through in-flight host discrimination instead of inspecting every single host whether it is parasitized or not by contact.

Future studies of these fascinating interactions between Brussels sprouts plants, *Pieris* butterflies, and their egg and larval parasitoids should extend the laboratory studies conducted in this thesis to natural environments. Field studies should reveal how butterfly anti-aphrodisiacs affect interactions between organisms of the different trophic levels. Natural variation in the production of anti-aphrodisiacs in *Pieris* butterflies and the correlation with parasitoid selection pressure should be examined to understand to what extent the espionage-and-ride strategy of *Trichogramma* wasps constrains the evolution of sexual communication between *Pieris* butterflies. Molecular and chemical studies on the signal transduction and the elicitor involved in the induction of unparasitized vs. parasitized *Pieris* caterpillar induced volatiles of Brussels sprouts plants are needed to evaluate the mechanism behind the differences in volatile bouquets.

References

Agelopoulos NG, Dicke M, Posthumus MA (1995) Role of volatile infochemicals emitted by feces of larvae in host-searching behavior of parasitoid *Cotesia rubecula* (Hymenoptera: Braconidae): a behavioral and chemical study. Journal of Chemical Ecology 21:1789-1811

Agelopoulos NG, Keller MA (1994) Plant-natural enemy association in the tritrophic system, *Cotesia rubecula-Pieris rapae*-Brassicaceae (Cruciferae): I. Sources of infochemicals. Journal of Chemical Ecology 20:1725-1734

Alborn HT, Turlings TCJ, Jones H, T., Stenhagen JG, Loughrin JH, Tumlinson JH (1997) An elicitor of plant volatiles from beet armyworm oral secretion. Science 276:945-949

Andersson J, Borg-Karlson A-K, Wiklund C (2000) Sexual cooperation and conflict in butterflies: a male-transferred anti-aphrodisiac reduces harassment of recently mated females. Proceedings of the Royal Society: Biological Sciences 267:1271-1275

Andersson J, Borg-Karlson A-K, Wiklund C (2003) Antiaphrodisiacs in Pierid butterflies: a theme with variation! Journal of Chemical Ecology 29:1489-1499

Arakaki N, Wakamura S, Yasuda T (1996) Phoretic egg parasitoid, *Telenomus euproctidis* (Hymenoptera: Scelionidae), uses sex pheromone of tussock moth *Euproctis taiwana*

(Lepidoptera: Lymantriidae) as a kairomone. Journal of Chemical Ecology 22:1079-1085

Beckage NE (1993) Games parasites play: The dynamic roles of proteins and peptides in the relationship between parasite and host. In: Beckage NE, Thompson SN, Federici BA (eds) Parasites and pathogens of insects, vol 1. Academic Press, San Diego, pp 25-57

Beckage NE, Kanost M, R. (1993) Effects of parasitism by the braconid wasp *Cotesia congregata* on host hemolymph proteins of the tobacco hornworm, *Manduca sexta*. Insect Biochemistry and Molecular Biology 23:643-653

Berdegue M, Trumble JT, Hare JD, Redak RA (1996) Is it enemy-free space? The evidence for terrestrial insects and freshwater arthropods. Ecological Entomology 21:203-217

Blaakmeer A, Geervliet JBF, van Loon JJA, Posthumus MA, van Beek TA, de Groot Æ (1994) Comparative headspace analysis of cabbage plants damaged by two species of *Pieris* caterpillars: consequences for in-flight host location by *Cotesia* parasitoids. Entomologia Experimentalis et Applicata 73:175-182

Colazza S, Fucarino A, Peri E, Salerno G, Conti E, Bin F (2004a) Insect oviposition induces volatile emission in herbaceous plants that attracts egg parasitoids. Journal of Experimental Biology 207:47-53

Colazza S, McElfresh JS, Millar JG (2004b) Identification of volatile synomones, induced by *Nezara viridula* feeding and oviposition on bean spp., that attract the egg parasitoid *Trissolcus basalis*. Journal of Chemical Ecology 30:945-964

Cortesero AM, Monge JP, Huignard J (1993) Response of the parasitoid *Eupelmus vuilleti* to the odors of the phytophagous host and its host-plant in an olfactometer. Entomologia Experimentalis et Applicata 69:109-116

De Moraes CM, Lewis WJ, Paré PW, Alborn HT, Tumlinson JH (1998) Herbivore-infested plants selectively attract parasitoids. Nature 393:570-573

De Moraes CM, Meschner MC, Tumlinson JH (2001) Caterpillar-induced nocturnal plant volatiles repel conspecific females. Nature 410:577-580

Denno RF, Larsson S, Olmstead K, L. (1990) Role of enemy-free space and plant quality in host-plant selection by willow beetles. Ecology 71:124-137

Dicke M (1994) Local and systemic production of volatile herbivore-induced terpenoids: their role in plant-carnivore mutualism. Journal of Plant Physiology 143:465-472

Dicke M (1999) Evolution of induced indirect defense of plants. In: Tollrian R, Harvell CD (eds) The Ecology and Evolution of Inducible Defences. Princeton University Press, Princeton, NJ, pp 62-88

Dicke M, Gols R, Ludeking D, Posthumus MA (1999) Jasmonic acid and herbivory differentially induce carnivore-attracting plant volatiles in lima bean plants. Journal of Chemical Ecology 25:1907-1922

Dicke M, Sabelis MW (1988) How plants obtain predatory mites as bodyguards. Netherlands Journal of Zoology 38:148-165

Dicke M, van Poecke RMP (2002) Signalling in plant-insect interactions: signal transduction in direct and indirect plant defense. In: Scheel D, Wasternack C (eds) Plant Signal Transduction. Oxford University Press, Oxford, pp 289-316

Dicke M, van Poecke RMP, de Boer JG (2003) Inducible indirect defence of plants: from mechanisms to ecological functions. Basic and Applied Ecology 4:27-42

Du YJ, Poppy GM, Powell W, Pickett JA, Wadhams LJ, Woodcock CM (1998) Identification of semiochemicals released during aphid feeding that attract parasitoid *Aphidius ervi*. Journal of Chemical Ecology 24:1355-1368

Fatouros NE, Bukovinszkine'Kiss G, Kalkers LA, Soler Gamborena R, Dicke M, Hilker M (2005a) Oviposition-induced plant cues: do they arrest *Trichogramma* wasps during host location? Entomologia Experimentalis et Applicata 115:207-215

Fatouros NE, Huigens ME, van Loon JJA, Dicke M, Hilker M (2005b) Chemical communication -

Butterfly anti-aphrodisiac lures parasitic wasps. Nature 433:704

Fatouros NE, Van Loon JJA, Hordijk KA, Smid HM, Dicke M (2005c) Herbivore-induced plant volatiles mediate in-flight host discrimination by parasitoids. Journal of Chemical Ecology 31:2033-2047

Frankie GW, Vinson SB, Coville RE (1980) Territorial behaviour of *Centris adani* and its reproductive function in the Costa Rican dry forest (Hymenoptera: Antrhophoridae). Journal of the Kansas Entomological Society 53:837-857

Gabrys BJ et al. (1997) Sex pheromone of cabbage aphid *Brevicoryne brassicae*: Identification and field trapping of male aphids and parasitoids. Journal of Chemical Ecology 23:1881-1890

Geervliet JBF, Vet LEM, Dicke M (1994) Volatiles from damaged plants as major cues in long-rage host-searching by the specialist parasitoid *Cotesia rubecula*. Entomologia Experimentalis et Applicata 73:289-297

Gilbert LE (1976) Postmating female odor in *Heliconius* butterflies: A male-contributed antiaphrodisiac? Science 193:419-420

Glinwood RT, Du YJ, Powell W (1999) Responses to aphid sex pheromones by the pea aphid parasitoids *Aphidius ervi* and *Aphidius eadyi*. Entomologia Experimentalis et Applicata 92: 227-232

Godfray HCJ (1994) Parasitoids - Behavioral and Evolutionary Ecology. Princeton University Press, Princeton, New Jersey

Gols R, Posthumus M, A., Dicke M (1999) Jasmonic acid induces the production of gerbera volatiles that attract the biological control agent *Phytoseiulus persimilis*. Entomol.Exp.Appl. 93:77-86

Gouinguené S, Alborn HT, Turlings TCJ (2003) Induction of volatile emissions in maize by different larval instars of *Spodoptera littoralis*. Journal of Chemical Ecology 29:145-162

Grasswitz TR, Paine TD (1992) Kairomonal effect of an aphid cornicle secretion on *Lysiphlebus testaceipes* (Cresson) (Hymenoptera, Aphidiidae). Journal of Insect Behavior 5:447-457

Gratton C, Welter S, C. (1999) Does "enemy-free space" exist? Experimental host shifts of an herbivorous fly. Ecology 80:773-785

Gross J, Fatouros NE, Neuvonen S, Hilker M (2004) The importance of specialist natural enemies for *Chrysomela lapponica* in pioneering a new host plant. Ecological Entomology 29:584-593

Happ GM (1969) Multiple sex pheromones of the mealworm beetle, *Tenebrio molitor* L. Nature 222: 180-181

Hilker M, Kobs C, Varama M, Schrank K (2002) Insect egg deposition induces *Pinus sylvestris* to attract egg parasitoids. Journal of Experimental Biology 205:455-461

Hilker M, Meiners T (2002) Induction of plant responses towards oviposition and feeding of herbivorous arthropods: a comparison. Entomologia Experimentalis et Applicata 104:181-192

Hilker M, Meiners T (2006) Early herbivore alert: Insect eggs induce plant defense. Journal of Chemical Ecology:in press

Hilker M, Stein C, Schröder R, Varama M, Mumm R (2005) Insect egg deposition induces defence responses in *Pinus sylvestris*: characterisation of the elicitor. Journal of Experimental Biology 208:1849-1854

Hoballah ME, Kollner TG, Degenhardt J, Turlings TCJ (2004) Costs of induced volatile production in maize. Oikos 105:168-180

Jones RL, Lewis WJ, Beroza M, Bierl BA, Sparks AN (1973) Host-seeking stimulants (kairomones) for the egg parasite, *Trichogramma evanescens*. Environmental Entomology 2:593-596

Kahl J, Siemens D, Aerts RJ (2000) Herbivore-induced ethylene suppresses a direct defense but not a putative indirect defense against adapted herbivore. Planta 210:336-342

Kennedy BH (1984) Effect of multilure and its components on parasites of *Scolytus multistriatus* (Coleoptera: Scolytidae). Journal of Chemical Ecology 10:373-385

Klijnstra JW (1986) The effect of an oviposition deterring pheromone on egglaying in *Pieris brassicae*.

Entomologia Experimentalis et Applicata 41:139-146

Kukuk P (1985) Evidence for an antiaphrodisiac in the sweat bee *Lasioglossum zephyrum* (Dialictus). Science 227:656-657

Laing J (1937) Host-finding by insect parasites. 1. Observations on the finding of hosts by *Alysia manducator*, *Mormoniella vitripennis* and *Trichogramma evanescens*. Journal of Animal Ecology 6:298-317

Lewis WJ, Jones RL, Nordlund DA, Gross JR. HR (1975) Kairomones and their use for mangament of entomophagous insects. II. Mechanisms causing increase in rate of parasitization by *Trichogramma* spp. Journal of Chemical Ecology 1:349-360

Lewis WJ, Jones RL, Sparks AN (1972) A host-seeking stimulant for the egg parasite, *Trichogramma evanescens*: its source and a demonstration of its laboratory and field activity. Annals of the Entomological Society of America 65:1087-1089

Lou YG, Ma B, Cheng JA (2005) Attraction of the parasitoid *Anagrus nilaparvatae* to rice volatiles induced by the rice brown planthopper *Nilaparvata lugens*. Journal of Chemical Ecology 31: 2357-2372

Loughrin JH, Manukian A, Heath RR, Turlings TCJ, Tumlinson JH (1994) Diurnal cycle of emission of induced volatile terpenoids by herbivore-injured cotton plants. Proceedings of the National Academy of Sciences of the United States of America 91:11836-11840

Mattiacci L, Dicke M (1995a) Host-age discrimination during host location by *Cotesia glomerata*, a larval parasitoid of *Pieris brassicae*. Entomologia Experimentalis et Applicata 76:37-48

Mattiacci L, Dicke M (1995b) The parsitoid *Cotesia glomerata* (Hymenoptera: Braconidae) discriminates between first and fifth larval instars of its host *Pieris brassicae*, on the basis of contact cues from frass, silk and herbivore-damaged leaf tissue. Journal of Insect Behavior 8:485-498

Mattiacci L, Dicke M, Posthumus MA (1994) Induction of parasitoid attracting synomone in Brussels sprouts plants by feeding of *Pieris brassicae* larvae: role of mechanical damage and herbivore elicitor. Journal of Chemical Ecology 20:2229-2247

Mattiacci L, Dicke M, Posthumus MA (1995) β-Glucosidase: An elicitor of the herbivore-induced plant odor that attracts host-searching parasitic wasps. Proceedings of the National Academy of Sciences of the United States of America 92:2036-2040

McCall PJ, Turlings TCJ, Lewis WJ, Tumlinson JH (1993) Role of plant volatiles in host location by the specialist parasitoid *Microplitis croceipes* Cresson (Braconidae: Hymenoptera). J.Insect Behav. 6:625-639

Meiners T, Hilker M (1997) Host location in *Oomyzus gallerucae* (Hymenoptera: Eulophidae), an egg parasitoid of the elm leaf beetle *Xanthogaleruca luteola* (Coleoptera: Chrysomelidae). Oecologia 112:87-93

Meiners T, Hilker M (2000) Induction of plant synomones by oviposition of a phytophagous insect. Journal of Chemical Ecology 26:221-232

Moraes MCB, Laumann R, Sujii ER, Pires C, Borges M (2005) Induced volatiles in soybean and pigeon pea plants artificially infested with the neotropical brown stink bug, *Euschistus heros*, and their effect on the egg parasitoid, Telenomus podisi. Entomologia Experimentalis et Applicata 115:227-237

Mumm R, Hilker M (2005) The significance of background odour for an egg parasitoid to detect plants with host eggs. Chemical Senses 30:337-343

Mumm R, Schrank K, Wegener R, Schulz S, Hilker M (2003) Chemical analysis of volatiles emitted by *Pinus sylvestris* after induction by insect oviposition. Journal of Chemical Ecology 29: 1235-1252

Noldus LPJJ, van Lenteren JC (1985) Kairomones for the egg parasite *Trichogramma evanescens* Westwood II. Effect of contact chemicals produced by two of its hosts, *Pieris brassicae* L. and *Pieris rapae* L. Journal of Chemical Ecology 11:793-800

Nordlund DA, Strand MR, Lewis WJ, Vinson SB (1987) Role of kairomones from host accessory gland secretion in host recognition by *Telenomus remus* and *Trichogramma pretiosum*, with partial characterization. Entomologia Experimentalis et Applicata 44:37-44

Nufio CR, Papaj DR (2001) Host marking behavior in phytophagous insects and parasitoids. Entomologia Experimentalis et Applicata 99:273-293

Ozawa R, Shiojiri K, Sabelis MW, Arimura GI, Nishioka T, Takabayashi J (2004) Corn plants treated with jasmonic acid attract more specialist parasitoids, thereby increasing parasitization of the common armyworm. Journal of Chemical Ecology 30:1797-1808

Paré PW, Tumlinson JH (1997) *De novo* biosynthesis of volatiles induced by insect herbivory in cotton plants. Plant Physiology 114:1161-1167

Pohnert G, Jung V, Haukioja E, Lempa K, Boland W (1999) New fatty acid amides from regurgitant of lepidopteran (Noctuidae, Geometridae) caterpillars. Tetrahedron 55:11275-11280

Potting RPJ, Vet LEM, Dicke M (1995) Host microhabitat location by stem-borer parasitoid *Cotesia flavipes*: the role of herbivore volatiles and locally and systemically induced plant volatiles. Journal of Chemical Ecology 21:525-539

Powell W (1999) Parasitoid hosts. In: Hardie J, Minks AK (eds) Pheromones of Non-Lepidopteran Insects Associated with Agricultural Plants. CABI Publishing, Wallingford, pp 405-427

Powell W, Pennacchio F, Poppy GM, Tremblay E (1998) Strategies involved in the location of hosts by the parasitoid *Aphidius ervi* (Hymenoptera: Braconidae: Aphidiinae). Biological Control 11:104-112

Prokopy RJ, Webster RP (1978) Oviposition-deterring pheromone of *Rhagoletis pomonella*, a kairomone for its parasitoid *Opius lectus*. Journal of Chemical Ecology 4:481-494

Röse USR, Manukian A, Heath RR, Tumlinson JH (1996) Volatiles semiochemicals released from undamaged cotton leaves: a systemic response of living plants to caterpillar damage. Plant Physiology 111:487-495

Rothschild M, Schoonhoven LM (1977) Assessment of egg load by *Pieris brassicae* (Lepidoptera: Pieridae). Nature 266:353-355

Rutledge CE (1996) A survey of identified kairomones and synomones used by insect parasitoids to locate and accept their hosts. Chemoecology 7:121-131

Schmelz EA, Alborn HT, Tumlinson JH (2001) The influence of intact-plant and excised-leaf bioassay designs of volicitin- and jasmonic acid-induced sesquiterpene volatile release in *Zea mays*. Planta 214:171-179

Scott D (1986) Sexual mimicry regulates the attractiveness of mated *Drosophila melanogaster* females. Proceedings of the National Academy of Sciences of the United States of America 83:8429-8433

Shu S, Jones RL (1989) Kinetic effect of a kairomone in moth scales of the European corn borer on *Trichogramma nubilale* Ertle & Davis (Hymenoptera: Trichogrammatidae). Journal of Insect Behavior 2:123-131

Smid HM, Van Loon JJA, Posthumus M, A., Vet LEM (2002) GC-EAG-analysis of volatiles from Brussels sprouts plants damaged by two species of *Pieris* caterpillars: olfactory respective range of a specialist and generalist parasitoid wasp species. Chemoecology 12:169-176

Steinberg S, Dicke M, Vet LEM, Wanningen R (1992) Response of the braconid parasitoid *Cotesia* (=*Apanteles*) *glomerata* to volatile infochemicals: effect of bioassay set-up, parasitoid age and experience and barometric flux. Entomologia Experimentalis et Applicata 63:163-175

Strand MR, Vinson SB (1982) Source and characterization of an egg recognition kairomone of *Telenomus heliothidis*; a parasitoid of *Heliothis virescens*. Physiological Entomology 7:83-90

Takabayashi J, Takahashi S, Dicke M, Posthumus MA (1995) Developmental stage of herbivore *Pseudaletia separata* affects production of herbivore-induced synomone by corn plants. Journal of Chemical Ecology 21:273-287

Takasu K, Hirose Y (1988) Host discrimination in the parasitoid *Ooencyrtus nezarae*: the role the egg stalk as an external marker. Entomologia Experimentalis et Applicata 47:47-48

Thaler JS (1999) Jasmonate-inducible plant defences cause increased parasitism of herbivores. Nature 399:686-688

Truitt CL, Wei HX, Pare PW (2004) A plasma membrane protein from Zea mays binds with the herbivore elicitor volicitin. Plant Cell 16:523-532

Turlings TCJ, Lengwiler UB, Bernasconi M, Wechsler D (1998) Timing of induced volatile emissions in maize seedlings. Planta 207:146-152

Turlings TCJ, McCall PJ, Alborn HT, Tumlinson JH (1993) An elicitor in caterpillar oral secretions that induces corn seedlings to emit chemical signals attractive to parasitic wasps. Journal of Chemical Ecology 19:411-425

Turlings TCJ, Tumlinson JH, Eller FJ, Lewis WJ (1991) Larval-damaged plants: source of volatile synomones that guide the parasitoid *Cotesia marginiventris* to the micro-habitat of its hosts. Entomologia Experimentalis et Applicata 58:75-82

Turlings TCJ, Tumlinson JH, Lewis WJ (1990) Exploitation of herbivore-induced plant odors by host-seeking parasitic wasps. Science 250:1251-1253

Turlings TCJ, Wäckers F (2004) Recruitment of predators and parasitoids by herbivore-injured plants. In: Carde R, Millar JG (eds) Advances in Insect Chemical Ecology. Cambridge University Press, Cambridge, pp 21-75

van Lenteren JC (1981) Host discrimination by parasitoids. In: Nordlund DA, Jones RL, Lewis WJ (eds) Semiochemicals- Their Role in Pest Control. John Wiley & Sons, New York, pp 153-179

Vet LEM, Dicke M (1992) Ecology of infochemical use by natural enemies in a tritrophic context. Annual Review of Entomology 37:141-172

Wegener R, Schulz S (2002) Identification and synthesis of homoterpenoids emitted from elm leaves after elicitation by beetle eggs. Tetrahedron 58:315-319

Wegener R, Schulz S, Meiners T, Hadwich K, Hilker M (2001) Analysis of volatiles induced by oviposition of elm leaf beetle *Xanthogaleruca luteola* on *Ulmus minor*. Journal of Chemical Ecology 27:499-515

Wiskerke JSC, Dicke M, Vet LEM (1993) Larval parasitoid uses aggregation pheromone of adult hosts in foraging behaviour: a solution to the reliability-detectability problem. Oecologia 93:145-148

Zaborski E, Teal PEA, Laing JE (1987) Kairomone-mediated host finding by spruce budworm egg parasite, *Trichogramma minutum*. Journal of Chemical Ecology 13:113-122

Zaki FN (1996) Effect of some kairomones and pheromones on two hymenopterous parasitoids *Apanteles ruficrus* and *Microplitis demolitor* (Hym, Braconidae). Journal of Applied Entomology-Zeitschrift Für Angewandte Entomologie 120:555-557

Zheng SJ, Bruinsma M, Dicke M (2006) Sensitivity and speed of induced defense of cabbage (*Brassica oleracea* L.): dynamics of *BoLOX* expression patterns during insect and pathogen attack.

10 Zusammenfassung

N.E. Fatouros

Chapter 10

Zusammenfassung

Wirtsfindungsstrategien bei Ei- und Larvalparasitoiden unter Ausnutzung chemischer Signale ihrer Wirte und Wirtspflanzen

Parasitoiden sind während der Wirtsfindung auf chemische Signale ihres Wirtes und seiner Umgebung angewiesen (Vinson 1991). Ziel dieser Dissertation ist es, neue Wirtsfindungsstrategien bei Ei- und Larvalparasitoiden, die chemische Signale der ersten und zweiten trophischen Ebene im Kohlweißlings-Rosenkohl System benutzen, zu entdecken.

Eiparasitoiden der Gattung *Trichogramma* sind weltweit verbreitet und sind auf verschiedene Arten von Schmetterlingen und Motten spezialisiert. Bisher konnte gezeigt werden, dass Trichogrammen vor allem Sexualduftstoffe der Imagines, sog. Sexualpheromone ausspionieren, um in die Nähe von Eiablageplätzen zu gelangen (Powell 1999). Wirtsflügelschuppen oder Düfte von unbefallen Wirtspflanzen können ebenso zur Anlockung dienen (Noldus 1989). Larvalparasitoiden, wie z.b. *Cotesia*-Arten, parasitieren vor allem die Raupen verschiedenster Schmetterlings- und Mottenarten. Zur Fernanlockung dienen ihnen hauptsächlich sog. fraßinduzierte Pflanzenduftstoffe (Agelopoulos and Keller 1994; Steinberg et al. 1992; Turlings et al. 1990). Pflanzen, die durch Herbivorenfraß beschädigt werden, produzieren daraufhin Duftsignale, mit denen sie die natürlichen Feinde der Herbivoren anlocken und damit indirekt bekämpfen (= indirekt induzierte Abwehr von Pflanzen gegen Herbivoren) (Dicke and Sabelis 1988; Turlings et al. 1990). Neuerdings konnte auch gezeigt werden, dass Pflanzen schon bevor die fraßschädigenden Larven schlüpfen, also noch während der Eiablage, Duftsignale produzieren mit denen Eiparasitoiden angelockt werden (Hilker and Meiners 2006; Hilker et al. 2002).

In meiner Dissertation habe ich mich hauptsächlich mit tritrophischen Interaktionen zwischen *Trichogramma* und *Cotesia* Arten (3. trophische Ebene) mit ihren herbivoren Wirt, dem Grossen Kohlweißling (*Pieris brassicae*) (2. trophische Ebene) und dessen Wirtspflanze, dem Rosenkohl (*Brassica oleraceae* var. *gemmifera*) beschäftigt (1. trophische Ebene) (s. Kapitel 1, Abbildung 1). Zunächst gebe ich als Einleitung im Kapitel 2 eine Übersicht über bisherige Erkenntnisse an Wirtsfindungsstrategien bei Eiparasitoiden im Verband mit chemischen Signalen. In den nachfolgenden Kapiteln werden dann die folgenden spezifischen Fragestellungen behandelt:

1. Frage: Spionieren *Trichogramma* Eiparasitoiden das sexuelle Kommunikationssytem von Kohlweißlingen aus, um an ihre Wirtseier zu gelangen?

Während der Paarung überreichen *Pieris brassicae* Männchen den Weibchen ein sog. Antiaphrodisiakum, Benzylcyanid, mit dessen Hilfe ihre Partnerinnen unattraktiver für konkurrierende Männchen werden (Andersson et al. 2003). Der Eiparasitoid *Trichogramma brassicae* kann sich dieses Wirtspheromon verpaarter Weibchen des Kohlweißlings zu nutze machen, indem er sie aufsucht, auf die Schmetterlinge aufsteigt, und zu ihren Eiablageplätzen mitfliegt (=Phoresie), um anschließend die frisch gelegten Eier zu parasitieren (Fatouros et al. 2005b) (Kapitel 3).

In Laborversuchen mit Hilfe eines sog. Zweikammerolfaktometers, wurde zunächst getestet, welche Schmetterlingsgerüche die Eiparasitoiden am meisten anlocken. Eindeutig bevorzugte *T. brassicae* die Düfte bereits gepaarter *P. brassicae* Weibchen sowie die von Männchen; jungfräuliche Weibchen wirkten dagegen nicht arretierend. Bei Applikation von Benzylcyanid, in zwei verschiedenen Konzentrationen (1 und 2 µg), auf jungfräuliche *P. brassicae* Weibchen, wirkten letztere allerdings arretierend auf den Eiparasitoiden gegenüber Weibchen, die nur mit dem Lösungsmittel behandelt wurden.

In weiteren Zweifachwahlversuchen in einer Arena sollte untersucht werden, ob die Eiparasitoiden das Antiaphrodisiakum auch dazu benutzen, um auf die verpaarten Schmetterlingsweibchen zu klettern, um sie dann als Transportmittel zu benutzen. *T. brassicae* kletterte bevorzugt auf verpaarte Kohlweißlingsweibchen im Vergleich zu Männchen oder unverpaarten Weibchen. Erst als unverpaarte *P. brassicae* Weibchen mit 2 µg Benzylcyanid behandelt wurden, und gegen unverpaarte Weibchen angeboten wurden, die nur mit dem Lösungsmittel behandelt waren, wurden erstere attraktiv für den Eiparasitoid und bevorzugt erklommen.

In Versuchen in einer Flugkammer sollte gezeigt werden, ob die Eiparasitoiden mit den verpaarten Schmetterlingsweibchen tatsächlich zur Wirtspflanze, dem Rosenkohl, mitfliegen und dann die frisch gelegten Wirtseier parasitieren. Von den insgesamt 28 getesteten verpaarten *P. brassicae* Weibchen, auf denen bei Abflug ein Eiparasitoid saß, trugen die Hälfte die Eiparasitoiden bis zur Landung auf der Wirtspflanze mit sich. Bei zwei Wespen konnte sogar beobachtet werden, wie sie vom Schmetterling herunterkletterten und die frisch gelegten Wirtseier parasitierten. Es konnte somit gezeigt werden, dass der Transport von *T. brassicae* auf verpaarten Schmetterlingsweibchen zu erfolgreicher Parasitierung in 7 % der beobachteten Eiparasitoiden führte.

Diese faszinierende Strategie, die Auffindung des Antiaphrodisiakums und „Trampen" auf dem Wirtstier, ist eine neue bei Trichogrammen bislang unbekannte Methode, um zu den Wirten zu gelangen. Wir erwarten jedoch, dass solche Art chemische Spionage in Kombination mit Phoresie zum Aufsuchen von Wirten sich möglicherweise häufiger entwickelt hat und unter Umständen die Evolution sexueller Kommunikation von Wirten einschränken kann.

2. Frage: Benutzen die Trichogrammen Duftstoffe des Rosenkohls, die durch die Eier des Kohlweißlings induziert werden?

Trichogrammen können sich auch sog. Pflanzensynomone zu nutze machen, um ihre Wirtseier zu finden, erst nachdem sie auf der Pflanze gelandet sind. Eier, die vor drei Tagen auf dem Rosenkohl abgelegt wurden, modifizieren die pflanzliche Oberfläche in einer Weise, dass *T. brassicae* arretiert wird und länger nach Eiern sucht, als auf unbelegten Pflanzen (Fatouros et al. 2005a) (Kapitel 4).

Tests im Olfaktometer ergaben, dass die Trichogrammen nicht auf Duftstoffe von mit *P. brassicae* Eiern belegten Rosenkohlpflanzen reagierten. Daraufhin wurden Zweifachwahltests mit kleinen Blattstückchen in Petrischalen durchgeführt und die Aufenthaltsdauer des Eiparasitoiden auf den verschieden behandelten Blattstückchen gemessen. Blattstückchen ohne Eier von Blättern, an denen sich drei Tage lang Eier befanden, wirkten hierbei arretierend auf die Wespen im Vergleich zu Blattstückchen von unbelegten Pflanzen. Drei Tage alte Eier werden von *T. brassicae* bevorzugt parasitiert. Blätter von Pflanzen, auf denen die Eier zwei Tage vorher abgelegt wurden, wirkten hingegen nicht arretierend. Einen Tag nach der Eiablage wirkten dagegen Schmetterlingsrückstände, wie z.B. die Flügelschuppen arretierend auf *T. brassicae*. Diese waren nach drei Tagen jedoch nicht mehr arretierend, so dass hier ein Einfluss von Wirtssignalen ausgeschlossen werden konnte. Die Befunde deuten also daraufhin, dass es drei Tage nach Eiablage zu chemischen Veränderungen in der Blattoberfläche des Rosenkohls kommt, die von den Trichogrammen nach der Landung wahrgenommen werden können.

In der Folge sollte in chemischen Analysen untersucht werden, welche Stoffe in der Rosenkohloberfläche für die Reaktion des Eiparasitoiden verantwortlich sein könnten (Kapitel 5). Zudem wurden molekulare Untersuchungen an zwei Genen, dem Lipoxygenase-Gen (*LOX*) und Myrosinase-Gen (*MYR*), die zwei Enzyme kodieren, die schon bekanntermaßen mit der induzierten Abwehr von Pflanzen zu tun haben, durchgeführt. In Oberflächenabwaschungen von eierbelegten Rosenkohlpflanzen konnte festgestellt werden, dass eines der für Kohlgewächse typischen Senföle, Indolacetonitril, in reduzierten Mengen vorkam verglichen mit unbelegten Pflanzen. Ob dieses Senföl dabei entscheidend für die Reaktion der Trichogrammen ist, muss in weiteren Untersuchungen festgestellt werden. Weiterhin gab es Indikationen für eine erhöhte Expression des MYR-Gens in eierbelegten Pflanzen im Vergleich von unbelegten Pflanzen.

Ebenso offen blieb nun die Frage des Elicitors: welcher Stoff in den Eiern oder rund um die Eier dafür verantwortlich ist, dass eine Induktion der Synomone im Rosenkohl drei Tage nach Eiablage stattfindet (Kapitel 6). Wieder wurden verschieden behandelte Blattstückchen in Petrischalen den Trichogrammen angeboten. Zunächst wurden Pflanzen mit den akzessorischen Drüsensekreten, (Sekrete die mit den Eiern abgegeben werden) von verpaarten oder jungfräulichen Kohlweißlingsweibchen behandelt. Dabei zeigte sich, dass Pflanzen, die mit dem Drüsensekret von verpaarten Wirtsweibchen behandelt wurden und gegenüber nur mit dem Lösungmittel behandelten Pflanzen getestet wurden, die *T. brassicae* arretierten – und zwar erst drei Tage nachdem die Sekrete auf die Pflanzen aufgetragen wurden. Mit dem Sekret von unverpaarten *P. brassicae* Weibchen behandelte Pflanzen wirkten dagegen nicht arretierend, auch nicht nach drei Tagen. Der Einfluss des bereits erwähnten Antiaphrodisiakums lag nun nahe. Tatsächlich konnten in chemischen Analysen der akzessorischen Drüsen Spuren (weniger als 1 ng) von Benzylcyanid (BC) gefunden werden und zwar nur in denen verpaarter Weibchen. In der Folge wurden verschiedene Konzentrationen von BC, aufgetragen auf die Pflanze, einen Tag und drei Tage nach Behandlung getestet. Hier wirkte nur eine höhere Konzentration, 200 µg, nach drei Tagen aber nicht nach einem Tag arretierend auf die Wespen. Erst wenn den Drüsen unverpaarter Weibchen geringe Mengen an BC, entweder 100 oder 1 ng zugefügt wurden, wurden die Eiparasitoiden drei Tage nach arretiert. Somit konnte deutlich gezeigt werden, dass das Antiaphrodisiakum in Verbindung mit dem Drüsensekret auch eine elicitierende Funktion hat, indem es Veränderungen in der Pflanzenoberfläche induziert, die auf den Eiparasitoiden arretierend wirken.

Die Männchen des Kohlweißlings „stellen sich und ihren Nachkömmlingen durch die Übergabe des Antiaphrodisiakums somit gleich zweimal ein Bein": Einmal indem der Benzylcyanidgeruch ihrer Partnerinnen von den spionierenden Trichogrammen direkt geortet wird, und in der Folge die frisch gelegten Eier parasitiert werden. Und ein zweites Mal, indem die Wirtsweibchen während der Eiablage etwas von dem überreichten Benzylcyanid unfreiwillig freigeben, das in der Folge die Pflanze zur indirekten Abwehrreaktion induziert, wodurch erneut Eiparasitoide arretiert werden.

Letztendlich konnte ich eine vollständige Abfolge eines Wirtsfindungsprozess einer Trichogrammenart im Labor aufdecken: zur Wirtsfindung können die Trichogrammen die Antisexualpheromone ihrer Wirte benutzen, um mit Hilfe ihrer Transportmittel, den verpaarten Schmetterlingsweibchen, zu den Wirtspflanzen mitzufliegen, wo sie dann, einmal auf der Pflanze gelandet, mit Hilfe induzierter Pflanzenstoffe an der Blattoberfläche weitere Eigelege auffinden können.

3. Frage: Benutzen *Cotesia* Parasitoiden durch Fraß induzierte Duftstoffe des Rosenkohls, um unparasitierte und parasitierte Raupen des Kohlweißlings von der Ferne diskriminieren zu können?

Die Fähigkeit, um bereits parasitierte Wirte von unparasitierten Wirten zu unterscheiden und erstere für die Eiablage zu meiden (= Wirtsdiskrimminie-rung), ist relativ verbreitet bei Parasitoiden (Lenteren van 1981). Eine erfolgreiche Fortpflanzung hängt stark von der Qualität und Größe des Wirtes ab. Dieser bildet nur eine begrenzte Nahrungsquelle und je mehr Parasitoidenlarven der selben Art sich diese teilen müssen (= Superparasitismus), desto mehr kann dies zu Konkurrenzkampf, reduzierter Fitness oder selbst Mortalität führen (Godfray 1994). Zur Vermeidung von Superparasitismus benutzen Parasitoidenweibchen Duftmarkierungen, indem sie nach der Eiablage den Wirt mit einem Pheromon markieren (Nufio and Papaj 2001). Weiterhin können auch Veränderungen der Wirtshämolymphe oder der Wirtsoberfläche nachfolgenden Parasitoidenweibchen signalisieren, ob der Wirt bereits belegt ist. Jedoch sind solche Markierungen meist nur über Kontakt oder in nächster Nähe des Wirtes feststellbar und meist auch nicht von großer Dauer.

In Windtunnel Experimenten sollte getestet werden, ob zwei verschiedene

Larvalparasitoiden, die gregäre *Cotesia glomerata* und die solitäre *C. rubecula* die Pflanzendüfte, an denen entweder parasitierte oder unparasitierte *Pieris*-Raupen fressen, im Flug diskrimmieren können (Fatouros et al. 2005c) (Kapitel 8). Da einmal parasitierte Raupen oft anders fressen als unparasitierte, wurden die Pflanzen mit dem sog. Regurgitat der Raupen behandelt und zusätzlich mechanischer Schaden beigefügt. Das Regurgitat plus mechanische Schädigung hat die gleiche induzierende Wirkung auf Pflanzen wie der Fraß selbst (Mattiacci et al. 1995).

Die Windtunnelstudien zeigten, dass sowohl *C. glomerata* als auch *C. rubecula* tatsächlich Unterschiede zwischen den verschieden induzierten Pflanzen wahrnehmen konnten und häufiger den Rosenkohl aufsuchten, der durch Regurgitat von unparasitierten Raupen behandelt wurde. Daraufhin wurden chemische Analysen der Pflanzendüfte durchgeführt. Hier zeigte sich, dass sich die Duftbouquets von Pflanzen behandelt mit Regurgitat von unparasitierten *P. brassicae* Raupen quantitativ von denen unterschieden, die mit Regurgitat von parasitierten *P. brassicae* Raupen behandelt waren. Insgesamt wurden die meisten Verbindungen in höheren Mengen von Pflanzen abgegeben, woran unparasitierte Raupen „fraßen", wobei v.a. drei Terpenoide signifikant mehr abgegeben wurden.

Auch diese Strategie der Wirtsdiskrimminierung mit Hilfe induzierter Pflanzenduftstoffe ist bisher unbekannt gewesen und kann sowohl für die Schlupfwespen als auch die Pflanzen sehr vorteilig sein. Die Schlupfwespen sparen Zeit und Energie, indem sie bereits von der Ferne erkennen können, ob sie es mit geeigneten Wirten zu tun haben. Die Pflanze kann biosynthetische und ökologische Kosten bei der Produktion solcher „SOS-Signale" sparen, indem sie ihre Produktion reduziert, nachdem die Herbivoren erfolgreich durch Anlockung von Parasitoiden verteidigt wurden.

In einer abschließenden Diskussion werden die verschiedenen, hier untersuchten Strategien von Larval- und Eiparasitoiden diskutiert, mit anderen tritrophischen Systemen verglichen und in einen breiteren ökologischen Kontext gesetzt.

Diese Dissertation gibt neue Einblicke in die faszinierende Welt der Interaktionen zwischen Pflanzen, herbivoren Insekten und parasitischen Wespen mittels chemischer Signale. Sie liefert weitere wichtige Grundlagen zum Verhalten von parasitischen Wespen und der Bedeutung von chemischen Duftstoffen zur erfolgreichen Wirtsfindung, die im gezielten und erfolgreichen Einsatz von

Parasitoiden gegen Schadinsekten auf Kulturpflanzen von größter Nützlichkeit sein können.

Literatur

Agelopoulos NG, Keller MA (1994) Plant-natural enemy association in the tritrophic system, *Cotesia rubecula-Pieris rapae*-Brassicaceae (Cruciferae): I. Sources of infochemicals. Journal of Chemical Ecology 20:1725-1734

Andersson J, Borg-Karlson A-K, Wiklund C (2003) Antiaphrodisiacs in Pierid butterflies: a theme with variation! Journal of Chemical Ecology 29:1489-1499

Dicke M, Sabelis MW (1988) How plants obtain predatory mites as bodyguards. Netherlands Journal of Zoology 38:148-165

Fatouros NE, Bukovinszkine'Kiss G, Kalkers LA, Soler Gamborena R, Dicke M, Hilker M (2005a) Oviposition-induced plant cues: do they arrest *Trichogramma* wasps during host location? Entomologia Experimentalis et Applicata 115:207-215

Fatouros NE, Huigens ME, van Loon JJA, Dicke M, Hilker M (2005b) Chemical communication - Butterfly anti-aphrodisiac lures parasitic wasps. Nature 433:704

Fatouros NE, Van Loon JJA, Hordijk KA, Smid HM, Dicke M (2005c) Herbivore-induced plant volatiles mediate in-flight host discrimination by parasitoids. Journal of Chemical Ecology 31:2033-2047

Godfray HCJ (1994) Parasitoids - Behavioral and Evolutionary Ecology. Princeton University Press, Princeton, New Jersey

Hilker M, Meiners T (2006) Early herbivore alert: insect eggs induce plant defense. Journal of Chemical Ecology:in press

Hilker M, Rohfritsch O, Meiners T (2002) The plant's response towards insect egg deposition. In: Hilker M, Meiners T (eds) Chemoecology of Insect Eggs and Egg Deposition. Blackwell Publishing, Berlin, Vienna, pp 205-233

Lenteren van JC (1981) Host discrimination by parasitoids. In: Nordlund DA, Jones RL, Lewis WJ (eds) Semiochemicals- Their Role in Pest Control. John Wiley & Sons, New York, pp 153-179

Mattiacci L, Dicke M, Posthumus MA (1995) β-Glucosidase: An elicitor of the herbivore-induced plant odor that attracts host-searching parasitic wasps. Proceedings of the National Academy of Sciences of the United States of America 92:2036-2040

Noldus LPJJ (1989) Semiochemicals, foraging behaviour and quality of entomophagous insects for biological control. Journal of Applied Entomology 108:425-451

Nufio CR, Papaj DR (2001) Host marking behavior in phytophagous insects and parasitoids. Entomologia Experimentalis et Applicata 99:273-293

Powell W (1999) Parasitoid hosts. In: Hardie J, Minks AK (eds) Pheromones of Non-Lepidopteran Insects Associated with Agricultural Plants. CABI Publishing, Wallingford, pp 405-427

Steinberg S, Dicke M, Vet LEM, Wanningen R (1992) Response of the braconid parasitoid *Cotesia* (=*Apanteles*) *glomerata* to volatile infochemicals: effect of bioassay set-up, parasitoid age and experience and barometric flux. Entomologia Experimentalis et Applicata 63:163-175

Turlings TCJ, Tumlinson JH, Lewis WJ (1990) Exploitation of herbivore-induced plant odors by host-seeking parasitic wasps. Science 250:1251-1253

Vinson SB (1991) Chemical signals used by parasitoids. Redia 74:15-42

Danksagung - Dankwoord - Acknowledgements

Diese Dissertation wäre ohne die Mithilfe vieler Personen in Berlin und Wageningen niemals entstanden. Darum möchte ich hiermit allen denjenigen danken, die mich in den letzten fünf Jahren mit begleitet und unterstützt haben.

First of all, I am very grateful to my two "doctor parents", Monika Hilker and Marcel Dicke: thanks to your fast and helpful comments on my manuscripts and chapters in the last couple of months, I was able to accomplish my thesis on time, though both of you were very busy and sometimes located at the other end of the world, like in Japan or Uganda. I enjoyed your supervision and our many meetings and learned a lot from you and your sometimes very inventive ideas. Ihnen, Frau Hilker, bin ich zudem sehr dankbar für die Überlassung des Themas, Ihr Vertrauen und Ihre Kulanz mich hauptsächlich „außerhalb Ihrer Reichweite" arbeiten zu lassen. Marcel, jij ook nog bedankt voor je geweldige ondersteuning en het mogelijk maken mij op Ento te laten werken.

Joop van Loon wil ik graag bedanken voor de vele, vaak spontane bijeenkomsten en leerzame discussies, het *Cotesia* project aan mij toe te vertrouwen en de vele, behulpzame commentaren op mijn manuscripten.

Gabriella, without the many hours you spent in the fitness room on the two-choice bioassays and on the rearing of "our" beautiful wasps, I would have finished a lot less work. Thanks a lot for all your help and also for your advices during the pregnancy period, all the best for you, Tibor and Villö!

Furthermore, I want to thank Si-Jun for the introduction into the molecular world and his patience with my not so "green fingers", Hans Smid for his help during the *Cotesia* work, the many "stupid" questions he had to endure and his technical advices/supports such as on macro photography, and Patrick for his support in the molecular lab and the entertaining hours with 3FM music.

Many thanks to Roxina and my one and only Msc-student Lukas Kalkers for their helpful work on the wasps' behavior.

I enjoyed the many roommates I had, thanks a lot Desire, Renate, Mohammed, Linnet, Trefi, Maartje, Valentina, Tibor and Gabriella for the "gezellige" discussions and hours we spent together in the "greenhouse corridor rooms" and for many nice evening activities. Many thanks to Tibor for some of his photos and for his advices, e.g., on statistics, photography or Adobe Photoshop© and the cozy atmosphere we have in our "pregnancy room".

Thank you William, Emmanuel and Desire for your open and joyful characters, and for showing me how different you African people sometimes live and think.

Frodo, Linde en Maartje bedankt voor de vele gezellige spelletjesavonden en etentjes and thank you Roxina, Marcos, Tibor and Gabriella for the nice dinner evenings.

Leo, Andre en Frans bedankt voor het dagelijkse kweekwerk, en Sabine voor het organizeren van de administratie op Ento. I enjoyed it being a member of the party committee together with Maaike, Sabine, Roland, Tibor, Ties, David, Niels, sometimes Yde and Peter, all creating a nice atmosphere. Thanks a lot to all members of Ento for a wonderful time in this "family culture".

Hannes und Roland, Euch halben Wageningern möchte ich besonders danken für die deutsche Unterstützung in einer doch nicht immer einfachen holländischen Gesellschaft. Hannes, danke für Deine Einführung in Wageningen, Deinen ansteckenden Enthusiasmus, die Vorversuche im 4-Arm Olfaktometer und vieles mehr. Roland, ich freue mich, dass es Dich nun sogar für längere Zeit nach Wageningen verschlagen hat. Danke für Deine Hilfe und Deine breiten Schultern bei diversen großen und kleinen Problemen und die vielen „gezelligen" Abende - zuletzt auch zusammen mit Sylvia.

Frank, vielen Dank für Deine Geduld und den enormen Aufwand mit den Blattextraktanalysen (und die vielen Male, die du das GC-MS deswegen putzen musstest).

Torsten und Joachim danke ich v.a. für viele Tips, hilfreiche Gespräche und ihren großartigen Humor. Ihr seid der lebende Beweis dafür, dass auch Deutsche Humor haben!

Jürgen danke ich dafür, dass er mich in die Welt der Wissenschaft mithineingeführt hat, sowie für die inzwischen vielen, gemeinsamen Manuskripte über „unsere" *C. lapponicas.*

Barbara möchte ich für ihre Mithilfe an den Phoresie Experimenten danken, Florian J., Tanja B., Johannes, Anna und Tanja für ihre Hiwi-Jobs. Urte, herzlichen Dank für Deine tolle Unterstützung in allen organisatorischen Belangen, Renate, Ute und allen anderen AG-Mitgliedern danke ich für die vielen netten Gespräche.

Also with members of both departments together, I have had great times on congresses, like with Maaike, Ties, Joachim, Florian, Sven and Dr. Boo at last year's ISCE meeting in Washington.

I am grateful to Meike Burow and Kees Hoordijk for their chemical analysis of cabbage plants.

Ich bedanke mich bei der Deutschen Forschungsgemeinschaft (DFG) für die Finanzierung eines Großteils dieser Arbeit (Projekt: Hi 416/15-1,2) und dem Land Berlin für ein Stipendium nach dem NaFöG.

Naast het werk waren er ook nog veel andere mensen die ik graag wil bedanken. Zonder mijn uitlaatventiel, het basketballen, zou ik veel vaker last van stress gehad hebben, dank aan alle basketbaldames in mijn ploegen bij Sphinx en Pluto voor de gezelligheid, en voor leuke en ontspannende sauna avonden met Liesbeth, Cornelly, Annemiek en vooral Marie-Claire, bedankt ook voor je dagelijkse bijkletsen via vrolijke e-mails.

I wanna thank my corridor mates Alena, Stephan, Martijn, Lucy, Cisca, Maaike, Tanya and many more for a nice time in Asserpark 6B-olé.

Mijn vrienden, buren en eetmaatjes Johan and Liesbeth: ik kon het van het begin af aan super goed met jullie vinden, bedankt dat jullie met mij, ondanks mijn Griekse buien, zo goed kunnen opschieten.

Während meines Studiums hatte ich viel Spass zusammen Liane, Nina K., Thilo, Alex und Miro, v.a. beim Öko-Kurs in Hüttenberg (Österreich).

Meinen verbliebenen Freunden aus Berliner Zeiten, Alex, Jule, Silke, Ilka, Ines, Christel, Daniel, Mira und Judith möchte ich für ihre treue Freundschaft danken, und dass sich das Sprichwort: „Aus dem Weg, aus dem Sinn" nicht immer bewahrheitet.

Aristea, many thanks for your true friendship since such a long time.

Liebe Mama, lieber Papa, danke für Euer Dasein, Euer Vertrauen und Eure Unterstützung in allen Belangen meines Lebens. Mama, danke, dass du mich jeden Tag auch gedanklich unterstützt, und dass du meine Wahl für Holland, trotz Distanz zu Berlin, so befürwortet hast. Auch den Rest der Deutsch-Griechischen Familie, meine großen Brüder Thomas und Nikos und Tante Käthoula sowie die angeheirateten PartnerInnen Barbara, Dietmar und Tina will ich hier natürlich nicht vergessen zu danken.

Sinds ruim viereneenhalf jaar heb ik ook een Nederlandse familie gevonden, die mij enorm ondersteunt. Dick, Coby, Erris, Heleen en de twee oma's wil ik hier in het bijzonder noemen. Jullie zijn voor mij al meer dan een schoonfamilie geworden. Coby en Dick, ik ben er zeker van dat jullie geweldige grootouders gaan worden.

Ja, en natuurlijk Ties, jou kan ik zeker niet vergeten te bedanken voor al je geduld, ondersteuning, liefde, positivisme, humor, en veel meer. Ik kan het geweldig met je vinden en ben er zeker van dat het straks met onze "hybride(n)" erbij alleen maar nog leuker gaat worden. Σ'αγαπω πολι!

Nina Fatouros, Juni 2006

Curriculum Vitae

Nina Katerina Eftichia Fatouros was born on a Sunday afternoon, March 17th 1974 in the Klinikum Steglitz of the Free University in (West-)Berlin (Germany). In 1980, she started her elementary school in the Rothenburg Grundschule and graduated from secondary school in the Fichtenberg Gymnasium (both in Berlin-Steglitz) in 1993. Inspired by her "Leistungskurs" Biology teacher she decided to follow a study in Biology at the Free University of Berlin. Soon she discovered her enthusiasm on the world of insects and their ecological interactions. After working on a short project on "The sex pheromone of the leaf beetle *Phaedon cochleariae*" at the Institute of Applied Ecology and Animal Ecology in 1998, she started her "Diplom" thesis on "The Evolution of host plant specialization of the leaf beetle *Chrysomela lapponica*" supervised by Dr. Jürgen Gross and Prof. Monika Hilker at the same institute in 1999. In two summer seasons (1999 & 2000), she carried out field work in the surroundings of the Kevo Subarctic Research Institute in North Lapland (Finland) of the University of Turku and finally graduated in November 2000. In June 2001, she was offered to work on a one-year project on "The host discrimination behavior of the larval parasitoids *Cotesia glomerata* and *C. rubelcula*" together with Dr. Joop van Loon and Prof. Marcel Dicke at the Laboratory of Entomology of the Wageningen University (The Netherlands). She got a PhD talent grant from the Berliner Graduierten Förderung (NaFöG) for two years to work on "The induction of Brussels sprouts plants by egg deposition of the large cabbage white in a tritrophic system" in July 2002. Later the same project was financed by the Deutsche Forschungsgemeinschaft (DFG) for 3 years. It was carried out at the Institute of Applied Ecology in Berlin and at the Laboratory of Entomology in Wageningen supervised by Prof. Monika Hilker and Prof. Marcel Dicke. The result of this 4-years PhD project is lying in front of you.

List of Publications

International (refereed) Journals

Gross, J., **Fatouros, N.E.**, & Hilker, M. (2004a) The significance of bottom-up effects for host plant specialization in *Chrysomela* leaf beetles. *Oikos*, **105**, 368-376.

Gross, J., **Fatouros, N.E.**, Neuvonen, S., & Hilker, M. (2004b) The importance of specialist natural enemies for *Chrysomela lapponica* in pioneering a new host plant. *Ecological Entomology*, **29**, 584-593.

Fatouros, N.E., Huigens, M.E., van Loon, J.J.A., Dicke, M., & Hilker, M. (2005a) Chemical communication - Butterfly anti-aphrodisiac lures parasitic wasps. *Nature*, **433**, 704.

Fatouros, N.E., Bukovinszkine'Kiss, G., Kalkers, L.A., Soler Gamborena, R., Dicke, M., & Hilker, M. (2005b) Oviposition-induced plant cues: do they arrest *Trichogramma* wasps during host location? *Entomologia Experimentalis et Applicata*, **115**, 207-215.

Fatouros, N.E., Van Loon, J.J.A., Hordijk, K.A., Smid, H.M., & Dicke, M. (2005c) Herbivore-induced plant volatiles mediate in-flight host discrimination by parasitoids. *Journal of Chemical Ecology*, **31**, 2033-2047.

Gross, J. & **Fatouros, N.E.** (2006) Striking differences in behaviour and ecology between some populations of *Chrysomela lapponica* L. *Bonner Zoologische Beiträge* (in press).

Fatouros, N.E., Bukovinszkine'Kiss, G., Dicke, M., & Hilker, H (2006a) The response specificity of *Trichogramma* egg parasitoids towards infochemicals during host location. *Journal of Insect Behavior* (submitted).

Fatouros, N.E., Hilker M., Gross J. (2006b) Reproductive isolation between populations from Northern and Central Europe of the leaf beetle *Chrysomela lapponica* L. *Chemoecology* (submitted).

National (edited) Journal

Fatouros, G. & **Fatouros, N.E.** (2001) Der Käfer in der altgriechischen Literatur. *Jahrbuch der Geschichte und Theorie der Biologie* **8**: 201-205.

Others

Fatouros, N.E. & Huigens M.E. (2006) Meeliftende sluipwespen zijn kindermoordenaars. In: *Muggenzifters en Mierenneukers: Insecten onder de Loep Genomen* (eds. M.E. Huigens & P.W. de Jong) (accepted).